Juran's Quality Essentials
For Leaders

Juran's Quality Essentials

For Leaders

Joseph A. DeFeo
Joseph M. Juran

New York   Chicago   San Francisco
Athens   London   Madrid
Mexico City   Milan   New Delhi
Singapore   Sydney   Toronto

Cataloging-in-Publication Data is on file with the Library of Congress.

McGraw-Hill Education books are available at special quantity discounts to use as premiums and sales promotions, or for use in corporate training programs. To contact a representative please visit the Contact Us page at www.mhprofessional.com.

Juran's Quality Essentials For Leaders

Copyright ©2014 by McGraw-Hill Education. All rights reserved. Printed in the United States of America. Except as permitted under the United States Copyright Act of 1976, no part of this publication may be reproduced or distributed in any form or by any means, or stored in a data base or retrieval system, without the prior written permission of the publisher.

1 2 3 4 5 6 7 8 9 0   DOC/DOC   1 9 8 7 6 5 4

ISBN   978-0-07-182591-7
MHID   0-07-182591-6

The pages within this book were printed on acid-free paper.

Sponsoring Editor	**Proofreader**
Judy Bass	Claire Splan
Acquisition Coordinator	**Indexer**
Amy Stonebraker	Jack Lewis
Editorial Supervisor	**Production Supervisor**
David E. Fogarty	Lynn M. Messina
Project Manager	**Composition**
Patricia Wallenburg	TypeWriting
Copy Editor	**Art Director, Cover**
James Madru	Jeff Weeks

Information contained in this work has been obtained by McGraw-Hill Education from sources believed to be reliable. However, neither McGraw-Hill Education nor its authors guarantee the accuracy or completeness of any information published herein, and neither McGraw-Hill Education nor its authors shall be responsible for any errors, omissions, or damages arising out of use of this information. This work is published with the understanding that McGraw-Hill Education and its authors are supplying information but are not attempting to render engineering or other professional services. If such services are required, the assistance of an appropriate professional should be sought.

About the Authors

JOSEPH M. JURAN was the international thought leader in the quality management field as the key driver of business performance for more than 70 years and is often referred to as "the father of modern-day quality management." He coined the universal concept of the "vital few and the useful many" Pareto Principle, which we know today as the 80–20 rule.

Dr. Juran became well known to the world after his first visit to Japan in 1954, soon after World War II. Professor Kano of Japan recalls, "He impressed top executives here with his managerial aspect of quality and contributed to the quality development in Japan by helping to establish the reputation of *made-in-Japan* products."

As an original member of the Board of Overseers, Dr. Juran helped to create the U.S. Malcolm Baldrige National Quality Award. In 1979, Dr. Juran founded the Juran Institute, an organization aimed at providing research and pragmatic solutions to enable organizations from any industry to learn the tools and techniques for managing quality.

The Juran Trilogy®, published in 1986, was accepted worldwide as the basis for quality management. After almost 50 years of research, his trilogy defined three management processes required by all organizations to improve: Quality control, quality improvement, and quality planning, which become synonymous with Juran and the Juran Institute, Inc.

As a result of the power and clarity of Joseph Juran's thinking and the scope of his influence, business leaders, legions of managers, and his fellow theorists worldwide recognize Dr. Juran as one of "the vital few"—a seminal figure in the development of management theory. Juran has contributed more to the field over a longer period of time than any other person, and yet, felt he had barely scratched the surface of his subject. "My job of contributing to the welfare of my fellow man," wrote Juran, "is the great unfinished business."

Dr. Juran is the author of more than 20 books.

JOSEPH A. DeFeo President and CEO of Juran Global, is recognized as one of the world's leading experts on transformational change models and breakthrough management principles. DeFeo has worked as a trusted adviser helping business leaders increase sales, reduce costs, and improve customer satisfaction through the deployment of performance excellence programs, including Business Process Management, Lean, Six Sigma, Strategic Planning, and Cultural Change Management. His ability to cut through complex issues and apply proven methodologies and solutions has made him a sought-after business partner for industry leaders globally.

Mr. DeFeo has ushered the Juran Institute into Juran Global to support a new age of business improvement by successfully building on the universal principles pioneered by Dr. Joseph Juran nearly 60 years ago while infusing current thinking and strategies around performance excellence and transformational change.

Mr. DeFeo is the co-author on *Quality Planning and Analysis* and also *Quality Management and Analysis,* with Dr. Frank Gryna; *Juran Institute's Six Sigma: Breakthrough and Beyond,* with his late mentor, Dr. William Barnard; and *Juran's Quality Handbook, 6th Edition: The Complete Guide to Performance Excellence,* the "go to" resource for deployment leaders with Dr. Joseph M. Juran.

CONTENTS

Preface xvii
Acknowledgments xix

CHAPTER 1 **Embrace Quality** 1
Superior Quality Always Leads to
    Sustainable Business Results 1
    Managing Quality Is Not Optional 2
Quality Impacts Revenue and Costs 4
Quality, Earnings, and the Stock Market. 5
Building Market Quality Leadership 5
Quality and Share of Market 6
    Effect of Quality Superiority 7
    No Quality Superiority 7
    Carryover of Failure-Prone Features 7
Changes in Customer Habits 8
The Twentieth Century and Quality 8
    Explosive Growth in Technology 9
    Threats to Human Safety, Health,
        and the Environment 9
    Consumer Movement 10
    Intensified International Competition in Quality .. 10
The Twenty-First Century and Quality 11
The Lessons Learned 11
References 13

CHAPTER 2 **Three Universal Quality Management Methods** 15
The Concept of Universals 15
    What Does Managing for Quality Mean? 16

　　　　　　　　　　Organizational Effectiveness Programs 20
　　　　　　　Management of Quality: The Financial
　　　　　　　　　and Cultural Benefits. 21
　　　　　　　　　　Features' Effect on Revenue 21
　　　　　　　　　　Failures' Effect on Income 21
　　　　　　　　　　Failures' Effect on Cost 22
　　　　　　　　　　How to Manage for Quality:
　　　　　　　　　　　A Financial Analogy 22
　　　　　　　The Juran Trilogy Diagram 24
　　　　　　　　　　Chronic and Sporadic 24
　　　　　　　　　　The Trilogy Diagram and Failures 26
　　　　　　　References 26

CHAPTER 3 **Leadership Role in Creating a Sustainable**
　　　　　　　Culture of Quality 27
　　　　　　　Culture Defined 27
　　　　　　　　　　What Does Culture Have to Do with
　　　　　　　　　　　Managing an Organization? 28
　　　　　　　　　　Transforming a Culture. 28
　　　　　　　Breakthrough and Transformational Change. 30
　　　　　　　　　　Breakthroughs Are Essential to
　　　　　　　　　　　Organizational Vitality. 30
　　　　　　　Breakthroughs in Leadership and Management 35
　　　　　　　Breakthroughs in Organizational Structure 37
　　　　　　　　　　Function-Based Organization 38
　　　　　　　　　　Business Process–Managed Organizations 39
　　　　　　　　　　Means of Achieving High Performance 39
　　　　　　　　　　Focus on External Customers. 40
　　　　　　　Breakthroughs in Current Performance 41
　　　　　　　Breakthroughs in Culture 42
　　　　　　　　　　How Are Norms Acquired? 43
　　　　　　　　　　How Are Norms Changed? 43
　　　　　　　Breakthroughs in Adaptability 49
　　　　　　　　　　The Route to Adaptability: The Adaptive Cycle
　　　　　　　　　　　and Its Prerequisites 50
　　　　　　　References 56

CHAPTER 4	**Aligning Quality Goals with the Strategic Plan** 57
	Strategic Planning and Quality: The Benefits......... 57
	What Is Strategic Planning and Deployment? 58
	Quality and Customer Loyalty Goals 60
	Why Strategic Deployment? The Benefits 61
	Why Strategic Deployment? The Risks........... 62
	Launching Strategic Planning and Deployment....... 63
	The Strategic Deployment Process 63
	Developing the Elements of Strategic Planning
	and Deployment............................... 66
	Establish a Vision........................... 66
	Agree on Your Mission....................... 68
	Develop Annual Goals....................... 70
	The Role of Leadership 73
	Subdivide and Deploy Goals.................. 74
	Deployment to Whom? 76
	A Useful Tool for Deployment 77
	Measure Progress with KPIs 77
	Reviewing Progress 80
	Competitive Quality............................. 81
	Performance on Improvement 81
	Costs of Poor Quality........................ 83
	Product and Process Failures 83
	Performance of Business Processes............. 83
	The Scorecard.............................. 84
	Business Audits............................. 85
	References 87
CHAPTER 5	**Product Innovation** 89
	Tackling the First Process of the Trilogy:
	Designing Innovative Products 89
	The Juran Quality by Design Model................. 90
	The Quality by Design Problem 93
	Juran Quality by Design Model 94
	Step 1—Establish: The Project and Design Goals .. 95
	Step 2—Define and Identify: The Customers 101

                Step 3—Discover: Customer Needs 103
                Step 4—Design: The Product or Service......... 112
                Step 5—Develop: The Process 122
                Step 6—Deliver: The Transfer to Operations..... 126
        References 127

CHAPTER 6   **Creating Breakthroughs in Performance**.......... 129
        The Universal Sequence for Breakthrough 129
                Unstructured Reduction of Chronic Waste 131
        Breakthrough Models and Methods 132
                Breakthrough Lessons Learned................ 133
                The Rate of Breakthrough Is Most Important 134
                All Breakthrough Takes Place
                    Project by Project......................... 136
                The Backlog of Breakthrough Projects
                    Is Never-Ending........................... 137
                Breakthrough Does Not Come Free 137
                Reduction in Chronic Waste Is Not
                    Capital-Intensive......................... 138
                The Return on Investment for Breakthrough
                    Improvement Is High 138
                The Major Gains Come from the Vital
                    Few Projects............................. 140
                Breakthrough—Some Inhibitors............... 141
                Disillusioned by the Failures.................. 141
                The Illusion of Delegation 141
                Employee Apprehensions..................... 142
        Securing Upper Management Approval
            and Participation 142
                Proof of the Need 143
                The Size of the Chronic Waste.................. 143
        COPQ versus Cost Reduction 144
                A Better Approach.......................... 146
        Driving Bottom-Line Performance 146
        The Results....................................... 148
                The Potential Return on Investment............ 148

Mobilizing for Breakthrough 151
   The Need for Formality 151
The Executive "Quality Council" 152
   Membership and Responsibilities 152
   Leaders Must Face Up to the Apprehensions
      about Elimination of Jobs 154
   Assistance from the Quality and/or
      Performance Excellence Functions 155
Breakthrough Goals in the Business Plan 156
   Deployment of Goals 156
   The Project Concept 157
   Use of the Pareto Principle 157
   The Useful Many Problems and Solutions 158
The Nomination and Selection Process 158
   Sources of Nominations 158
   Criteria for Projects 159
   Project Selection 160
   Vital Few and Useful Many 161
   Cost Figures for Projects 162
   Costs versus Percentage of Deficiencies 162
   Elephant-Sized and Bite-Sized Projects 163
   Replication and Cloning 163
   Model of the Infrastructure 164
Team Organization 165
   The Team Leader 165
   The Team Members 166
   Finding the Time to Work on Projects 166
   Facilitators and Black Belts 167
   The Qualifications of Facilitators and
      Black Belts 169
   Leaders Must Learn Key Breakthrough
      Terminology 171
   Diagnosis Should Precede Remedy 171
   Institutionalizing Breakthrough 172
Review Progress 172
References 173

CHAPTER 7   Assuring Repeatable and Compliant Processes 175

Compliance and Control Defined 175
    The Relation to Quality Assurance............. 177
    The Feedback Loop........................... 178
The Elements of the Feedback Loop 180
    The Control Subjects 180
    Establish Measurement 181
    Establish Standards of Performance:
        Product Goals and Process Goals 182
    Measure Actual Performance 184
    The Sensor................................... 184
    Compare to Standards........................ 185
    Take Action on the Difference................. 185
    The Key Process 185
    Taking Corrective Action..................... 186
The Pyramid of Control........................... 187
    Control by Technology (Nonhuman Means)..... 188
    Control by the Employees (Workforce).......... 189
    Control by the Managerial Hierarchy 189
Planning for Control 190
    Critical to Quality (CTQ): Customers
        and Their Needs.......................... 190
    Compliance and Control Concepts.............. 191
    Process Capability 191
Process Conformance 192
    Special and Common Causes of Variation 192
    The Shewhart Control Chart 192
    Points Within Control Limits 193
    Points Outside of Control Limits 194
    Statistical Control Limits and Tolerances 195
    Self-Control and Controllability............... 197
    Effect on the Process Conformance Decision..... 198
Product Conformance: Fitness for Purpose......... 199
    The Product Conformance Decision 199
    Self-Inspection 200
    The Fitness for Purpose Decision 201

	Disposition of Unfit Product.................202
	Corrective Action........................203
	Diagnosing Sporadic Change...............204
	Corrective Action—Remedy................205
	The Role of Statistical Methods in Control..........205
	Statistical Process Control (SPC).............205
	The Merits..............................206
	The Risks...............................206
	Information for Decision Making.............207
	The Quality Control System and Policy Manual.....208
	Provision for Audits............................209
	Tasks for Leaders..............................209
	References....................................210
CHAPTER 8	**Simplifying Macro Processes with Business Process Management**..........................211
	Why Business Process Management?..............211
	The BPM Methodology.........................214
	Deploying BPM...........................215
	Organizing for BPM.......................216
	Establishing the Team's Mission and Goals......217
	The Planning Phase: Planning the New Process.....217
	Defining the Current Process................218
	Discovering Customer Needs and Mapping the Current State.......................219
	Establishing Process Measurements...........221
	Analyzing the Process......................225
	Redesigning the Process....................226
	The Transfer Phase: Transferring the New Process Plan to Operations....................229
	Planning for Implementation Problems........229
	Creating Readiness for Change...................231
	Planning for Implementation Action..........232
	Deploying the New Process Plan..............232
	Operational Management Phase: Managing the New Process.....................233

   Business Process Metrics and Control 233
   Business Process Improvement 234
   Periodic Process Review and Assessment. 234
  The Future of BPM Combined with Technology 234
  References 235

CHAPTER 9 **Benchmarking to Sustain Market Leadership...... 237**
  Benchmarking: What It Is and What It Is Not 237
  Objectives of Benchmarking. 240
  Why Benchmark? 241
  Classifying Benchmarking 241
   Subject Matter and Scope (What) 242
   Internal and External, Competitive
    and Noncompetitive Benchmarking (Who) ... 245
   Data and Information Sources (How) 247
   Database Benchmarking. 248
   Survey Benchmarking 249
   Self-Assessment Benchmarking. 249
   One-to-One Benchmarking. 250
   Consortium Benchmarking. 250
  Benchmarking and Designing New Products 251
  Benchmarking and Long- and Short-Term Planning . 252
  The Benchmarking Process 253
  Planning and Project Setup. 255
  Data Collection and Normalization. 256
   Participant Support During the Benchmarking
    Process 256
   Data Validation 256
   Data Normalization 258
  Analysis and Identification of Best Practices 259
  Report Development 260
  Learning from Best Practices 260
   Internal Forums 260
   One-to-One Benchmarking. 261
   Best Practice Forums 261
  Improvement Action Planning and Implementation .. 262

Institutionalizing Learning 264
Legal and Ethical Aspects of Benchmarking 264
    The Benchmarking Code of Conduct 264
    Confidentiality 265
Managing for Effective Benchmarking 266
References 266

Index 267

PREFACE

Another book about quality—what a novel idea! There have been hundreds of books about quality since I began my career in 1983 as a facilitator of improvement teams at the PerkinElmer Corporation. There probably have been thousands written since Dr. Juran began his career in the 1930s. So why another book about quality? It is simple; because it works. Without a focus on quality as seen by the customers, there would not be Six Sigma of the last decade or even Lean methods of the present day. These methods and others that will follow will exist only because the purpose of an organization is to make money or meet budgets by satisfying the needs of its customers. Customers demand better quality of products and services at a better cost. The customer always needs more for less. Quality pays and quality works! Leaders must understand and embrace quality.

I have been working around the world for almost 30 years and I have seen leaders state they "embraced quality" as a business strategy and succeeded at making their companies the best in the land. I have also seen those that "embraced quality" and did not succeed and soon went out of business. Why two different stories?

The lessons learned from my experiences were similar to that of Dr. Juran. Dr. Juran wrote about the lessons learned for most of his 70-year career. The biggest lessons learned are that no organization will succeed to sustain success by ignoring the customer and their needs. An organization turns needs into products and if they are great, results in positive financial outcomes and if they do not, the opposite happens. Financial losses abound.

Understanding how to use quality as a strategic tool pays off. It has a great ROI compared to other shorter-term methods. A Juran book on quality is not about design of experiments or advanced statistics. A Juran text is about universal principles that once "embraced" will lead to finding the right tools and organization results. Embracing quality is not difficult but it does require a different way of thinking. Once grasped, it will lead to driving business performance, operational excellence, Lean-ness, zero defects, and financial results.

The nine chapters in this book represent the essentials of modern day business performance management driven by quality. Managing quality of products, processes, and people will lead to profit—always. These chapters are based on the successful *Juran Quality Handbook*, currently in print for a seventh decade! The chapters have evolved over time but the essence of Dr. Juran is still present. Why? The same reason 2 + 2 always = 4. It is a universal principle. Universals stay with us a long time. This book is about the universal principles of managing for quality. They are essential to success and must be essential for leaders. This is why I wrote another book on quality—to get a new generation of leaders to learn from the best. Enjoy the guidance.

ACKNOWLEDGMENTS

Writing and editing a text following in the footsteps of a legend like Dr. Juran is not easy. It also cannot be done alone. There are many people in my organization and my business life that I would like to recognize for their support for this book, but there are too many to mention by name. A few have been vital and should be acknowledged. Let me begin with Dr. Juran.

I would like to thank **Dr. Joseph M. Juran**, founder of Juran Institute, Inc., for his many contributions to the field of quality management including the basis for five of these chapters. Having the chance to work with Dr. Juran for almost 20 years was a blessing. His clear pragmatic advice to leaders was always welcomed and listened to. It is the greatest takeaway I have learned from working with Dr. Juran. If I can provide pragmatic solutions to leaders, there is a greater chance of success.

R. Kevin Caldwell, Executive Vice President, Juran Global, has been a loyal practitioner for Juran Global for over 15 years. His background as a Lean sensei and quality expert are apparent in Chapter 7: "Assuring Repeatable and Compliant Processes." To carry out the universal principle of quality control would not happen at Juran without the knowledge of Kevin. He is a true zealot of quality in a Lean world. Thank you for your time contributing to this chapter.

Brad Wood, Ph.D., International Managing Director, Juran Europe, contributed to Chapter 9: "Benchmarking to Sustain Market Leadership." His life's work has been engaged in benchmarking the best to share with the rest. Bringing a European perspective to our culture and this book enables us to meet the needs of our global supporters.

Joseph M. DeFeo, Director of Operations, Juran Global, and **Janice Doucet Thompson**, Past Director, Organizational Effectiveness, Sutter Health, contributed to Chapter 3: "Leadership Role in Creating a Sustainable Culture of Quality." An organization cannot change unless it knows what the baseline is and what you want to change into. This chapter and the experiences of these key authors provide a new look at an old subject—a quality culture.

Lastly, this book would not have happened without the Juran and McGraw-Hill teams.

Special thanks to Tina Pietraszkiewicz, my Assistant and Associate Editor, who made this book happen on time. Her persistence in getting me to complete my edits made it all happen.

To the Juran team and partners, a heartfelt thank you for your contributions to carrying out the mission of Dr. Juran: Jeremy Hopfer, Michelle Matschke, Audra D'Agostino, Peter Robustelli, Kaitlin Tyer, Mary Beth Edmond, Tracey King, Jonathan Flanders, Er Ralston, Dennis Monroe, John Early, Ian Fairbairn, Aideen McCrave, Brian Stockhoff, Adriaan du Plessis, Ruedi Bachmann, Mike Moscynski, Ryan Walker, Chuck Aubrey, Tom Casey, and Dr. David Fearon.

Juran's Quality Essentials
For Leaders

CHAPTER 1

Embrace Quality

> Organizations that engage in the relentless pursuit of delivering high-quality goods and services outperform those that do not.

This chapter clarifies the impact that the *quality* of products, services, and processes has on business performance. A business that has products that are superior to its competitors' in quality will outperform those that do not. Any organization can achieve measurable business results through the application of the universal methods to manage for quality. These methods include the design of quality, the control of quality, and the methods to continually improve the quality of goods, services, and processes. Leaders must not think that managing quality is a "fad" or that "we already did that." The management of quality is as important as managing for finance.

Superior Quality Always Leads to Sustainable Business Results

Superior *quality* goods and services will result in sustainable financial results because goods and services that are superior to the competitors' are salable. Goods and services that are salable because of quality continually drive revenue and maintain lower costs, leading to greater profitability. The pursuit of superior quality will transform the business and create a favorable quality culture.

Transformational change does not happen haphazardly. Superiority in quality from a customer perspective does not just happen. The business must make quality happen. Quality happens when the organization sets the strategic direction with a relentless pursuit to be the best in quality.

Organizations that attain superior results by designing, controlling, and continuously improving the quality of their goods and services are often called *world-class* or *vanguard companies*. They have achieved a state of performance excellence. Organizations that have attained a state of superior quality are well respected by customers because their products and services exceed customers' expectations, which leads to sustainable business results.

This pursuit of excellence through quality management methods creates greater customer, stakeholder, and employee satisfaction, which enables the organization to sustain performance over a longer term.

Managing Quality Is Not Optional

A common argument occurs among managers when asked, Does high quality cost more, or does high quality cost less? Seemingly they disagree. One-half agree that it costs more, and the other half feel it costs less. They are both right. The culprit is the word *quality*. It is spelled the same way and pronounced the same way, but has multiple meanings. To manage for superior quality and results, leaders must define the word *quality* from the perspective of customers—those people who buy the goods, services, and even the reputation of your organization.

At one financial services company, the leaders would not support a proposal to reduce wasteful business processes because the staff had labeled them *quality improvement*. Some of the leaders felt improving quality would cost more money. In their view, higher quality meant higher cost. Others felt it would cost less. The subordinates were forced to rename the proposal *productivity improvement* to secure approval and avoid confusion. Such confusion can be reduced if each organization makes clear the distinction between the multiple meanings of the word *quality*. However, some confusion is inevitable as long as we use a single word to convey very different ideas.

Leaders must have a common understanding of quality so they can manage it. First, agree on the meaning of the word *quality* as it applies to your business and its customers. Once defined, then it can be managed. If it can be managed, then it can be provided to the satisfaction of customers and stakeholders. Without a common understanding of the word *quality*, the organization will continue to make many short-term initiatives to improve quality and it will lead to "initiative overload."

There have been efforts to clarify matters by adding supplemental words. There also have been efforts to coin a short phrase that would clearly and simultaneously define both major meanings of the word *quality*. A popular definition was first presented in *Juran's Quality Handbook*. Quality was defined as meaning "fitness for use." Dr. Deming used "conformance to requirements." Robert Galvin, Chairman Emeritus of Motorola, used *Six Sigma* to distinguish the high level of quality as it related to defects. Others stated that quality means world-class excellence or best-in-class and performance excellence.

We have settled on quality as *fitness for purpose*. The purpose is defined by the customer's needs. These needs drive the purchase of your goods and services. If an organization understands the needs of its many customers, it should be able to design goods and services that are fit for purpose. No matter what the organization produces—a good or a service—it must be fit for its purpose. To be fit for purpose, every good, service, and interaction with customers must have the right features (characteristics of the good or service that satisfies customer needs) and be free of failure.

It is unlikely that any short phrase can provide the depth of meaning needed by leaders and managers who are faced with choosing a course of action to improve quality. The best you can do is to understand the distinctions set out in Table 1.1 and define quality based on these distinctions.

Table 1.1 The Meaning of Quality (From Juran and DeFeo, 2010)

Features That Meet Customer Needs	Freedom from Failures
Higher quality enables organizations to:	Higher quality enables organizations to:
• Increase customer satisfaction	• Reduce error rates
• Make products salable	• Reduce rework, waste
• Meet competition	• Reduce field failures, warranty charges
• Increase market share	• Reduce customer dissatisfaction
• Provide sales income	• Reduce inspection, test
• Secure premium prices	• Shorten time to put new products on the market
• Reduce risk	• Increase yields, capacity
	• Improve delivery performance
Major effect is on revenue.	Major effect is on costs.
Higher quality costs more.	Higher quality costs less.

Table 1.1 presents two of the many meanings of the word *quality*. These two are of critical importance to managing for quality.

Quality Impacts Revenue and Costs

First, quality has a big effect on *costs*. In this case, *quality* has come to mean freedom from troubles traceable to office errors, factory defects, field failures, and so on. *Higher quality* means fewer errors, fewer defects, and fewer field failures. When customers perceive a service or good as low-quality, they usually refer to the failures, the defects, the poor response times, etc.

To increase this type of quality, an organization must master the universal of quality improvement. This is often called *breakthrough* or *Six Sigma*. It is a systematic method to reduce the number of such deficiencies or the "costs of poor quality" to create a greater level of quality and fewer costs related to it.

Second, quality has an effect on *revenue*. In this case, *higher quality* means delivery of those features of the good or service that respond better to customer needs. Such features make the product or service salable. Since the customers value the higher quality, they buy it and you get revenue from it. It is well documented that being the quality leader can also generate premium prices and greater revenue.

The effects on costs and on revenue interact with one another. Not only do goods or services with deficiencies add to suppliers' and customers' costs, but also they discourage repeat sales. Customers who are affected by field failures are, of course, less willing to buy again from the guilty supplier. In addition, such customers do not keep this information to themselves—they publicize it so that it affects the decisions of other potential buyers, with negative effects on the sales revenue of the supplier.

The effect of poor quality on organizational finances has been studied broadly. In contrast, study of the effect of quality on revenue has lagged. This imbalance is even more surprising, since most upper managers give higher priority to increasing revenues than to reducing costs. This same imbalance presents an opportunity for improving organization economics through better understanding of the effect of quality on revenue.

Quality, Earnings, and the Stock Market

At the most senior levels of management and among board members, there is keen interest in financial metrics such as net income and share price. It is clear that different levels of quality can greatly affect these metrics, but so do other variables. Variables such as market choices, pricing, and financial policy can influence these metrics. Separating out the market benefits of managing for quality has just become feasible.

During the early 1990s, some of the financial press published articles questioning the merits of the Malcolm Baldrige National Quality Award, Six Sigma, and other similar initiatives to improve performance. These articles were challenged with an analysis of the stock price performance of organizations known to practice these methods. The Baldrige winners were compared to the performance of the S&P 500 as a whole. The results were striking. The Malcolm Baldrige National Quality Award winners outperformed the S&P 500. The Baldrige winners had advanced 89 percent, as compared to only 33 percent for the broad Standard & Poor's Index of 500 stocks ("Betting to Win on the Baldie Winners," 1993, p. 8). This set of winners became known as the "Baldie Fund."

The impact of the quality universals is also clear for organizations that are not measured by the performance of their asset values. Michael Levinson, City Manager of 2007 Award Recipient for the City of Coral Springs, stated it this way: "People ask, 'Why Baldrige?' My answer is very simple: Triple A bond rating on Wall Street from all three ratings agencies, bringing capital projects in on time and within budget, a 96 percent business satisfaction rating, a 94 percent resident satisfaction rating, an overall quality rating of 95 percent, and an employee satisfaction rating of 97 percent . . . that's why we're involved with Baldrige."

Building Market Quality Leadership

Market quality leadership is often the result of entering a new market first and gaining superiority. However, once superiority is gained, it must be maintained through continuing product or service improvements, or it could be lost if another organization decides to redefine that market by improving its quality. That supplier will gain superiority over the market leader and will become the *quality leader*. Organizations that have attained this leadership have usually done so on the basis of a strategic choice. They adopted a positive strategy to establish leadership as a formal business goal and then set out the means to reach that goal. Once attained, quality leadership endures until there is clear cumulative evidence that some competitor has overtaken the leadership. Lacking such evidence, the leadership can endure for decades and even centuries. However, superior quality can also be lost through some catastrophic change.

The growth of competition in quality has stimulated the expansion of strategic business planning to include planning for quality and quality leadership.

Quality and Share of Market

Growth in market share is often among the highest goals of upper managers. Greater market share means higher sales volume. In turn, higher sales volume accelerates return on investment disproportionally due to the workings of the break-even chart.

In Figure 1.1, to the right of the break-even line, an increase of 20 percent in sales creates an increase of 50 percent in profit, since the fixed costs

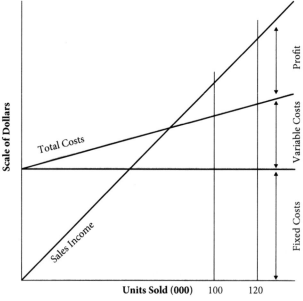

Figure 1.1 Break-even chart. (From Juran, 1999, p. 7.13)

do not increase. (Actually, constant costs do vary with volume, but not at all in proportion.) The risks involved in increasing market share are modest, since the technology, product or service, facilities, market, and so on are already in existence and of proved effectiveness.

Effect of Quality Superiority

Quality superiority can often be translated into higher share of market, but it always requires a special effort to do so. Superior quality must be clearly based on the voice of the customer and the benefits the customer is seeking. If quality superiority is defined only in terms of the company's internal standards, the customer may not perceive the value. For example, patients may be willing to pay the extra cost to travel long distances to a top health care system such as the Mayo Clinic in the United States rather than visit a local hospital because they perceive the superior clinical outcomes available at the Mayo Clinic.

No Quality Superiority

If there is no demonstrable quality superiority, then share of market is determined by marketing skills. These take such forms as persuasive value propositions, attractive packaging, and so on. Price reductions in various forms can provide increases in share of market, but this is usually temporary. Competitors move promptly to take similar action. Such price reduction can have a permanent effect if the underlying cost of production has also been reduced as the result of process improvements that give the company a competitive cost edge over its competitors.

Carryover of Failure-Prone Features

Market leadership can be lost by perpetuating failure-prone features of predecessor models. The guilty features are well known, since the resulting field failures keep the field service force busy restoring service. Nevertheless, there has been much carryover of failure-prone features into new models. At the least, such carryover perpetuates a sales detriment and a cost burden. At its worst, it is a cancer that can destroy seemingly healthy product or service lines.

A notorious example was the original xerographic copier. In that case. the "top 10" list of field failure modes remained essentially identical, model after model. A similar phenomenon existed for years in the automobile industry.

The reasons behind this carryover have much in common with the chronic internal wastes that abound in so many organizations:

1. The alarm signals are disconnected. When wastes continue, year after year, the accountants incorporate them into the budgets. That disconnects the alarm signals—no alarms ring as long as actual waste does not exceed budgeted waste.
2. There is no clear responsibility to get rid of the wastes. There are other reasons as well. The technologists have the capability to eliminate much of the carryover. However, those technologists are usually under intense pressure from the marketers to develop new product or service and process features in order to increase sales. In addition, they share distaste for spending their time cleaning up old problems. In their culture, the greatest prestige comes from developing the new.

The surprising result can be that each department is carrying out its assigned responsibilities, and yet the product or service line is dying. Seemingly nothing short of upper management intervention—setting goals for getting rid of the carryover—can break up the impasse.

Changes in Customer Habits

Customer habits can be notoriously fickle. Obvious examples are fashions in clothing or concerns over health. Consumerism is now driving lifestyles. Many people have reduced their consumption of beef and increased that of poultry and fish. Such shifts are not limited to consumers. Industrial organizations often launch "drives," most of which briefly take center stage and then fade away. The associated buzzwords similarly come and go.

The Twentieth Century and Quality

The twentieth century witnessed the emergence of some massive new forces that required responsive action. These forces included an explosive growth in science and technology; threats to human safety, health, and the environment; the rise of the consumerism movement; and intensified international competition in quality.

Explosive Growth in Technology

This growth made possible an outpouring of numerous benefits to human societies: longer life spans, superior communication and transport, reduced household drudgery, new forms of education and entertainment, and so on. Huge new industries emerged to translate the new technology into these benefits. Nations that accepted industrialization found it possible to improve their economies and the well-being of their citizenry.

The new technologies required complex designs and precise execution. The empirical methods of earlier centuries were unable to provide appropriate product and process designs, so process yields were low and field failures high. Organizations tried to deal with low yields by adding inspections to separate the good from the bad. They tried to deal with field failures through warranties and customer service. These solutions were costly, and they did not reduce customer dissatisfaction. The need was to prevent defects and field failures from happening in the first place.

Threats to Human Safety, Health, and the Environment

With benefits from technology came uninvited guests. To accept the benefits required changes in lifestyle, which, in turn, made quality of life dependent on continuity of service. However, many products were failure-prone, resulting in many service interruptions. Most of these were minor, but some were serious and even frightening—threats to human safety and health as well as to the environment.

Thus the critical need became quality. Continuity of the benefits of technology depended on the quality of the goods and services that provided those benefits. The frequency and severity of the interruptions also depended on quality—on the continuing performance and good behavior of the products of technology. This dependence came to be known as "life behind the quality dikes."

Consumer Movement

Consumers welcomed the features offered by the new products but not the associated new quality problems. The new products were unfamiliar—most consumers lacked expertise in technology. Their senses were unable to judge which of the competing products to buy, and the claims of competing organizations often were contradictory.

When products failed in service, consumers were frustrated by vague warranties and poor service. "The system" seemed unable to provide recourse when things failed. Individual consumers were unable to fight the system, but collectively they were numerous and hence potentially powerful, both economically and politically. During the twentieth century, a "consumerism" movement emerged to make this potential a reality and to help consumers deal more effectively with these problems. This same movement also was successful in stimulating new government legislation for consumer protection. (For elaboration, see Juran, *Managerial Breakthrough*, 1995, Chapter 17.)

Intensified International Competition in Quality

Cities and countries have competed for centuries. The oldest form of such competition was probably in military weaponry. This competition then intensified during the twentieth century under the pressures of two world wars. It led to the development of new and terrible weapons of mass destruction.

A further stimulus to competition came from the rise of multinational organizations. Large organizations had found that foreign trade barriers were obstacles to export of their products. To get around these barriers, many set up foreign subsidiaries that then became their bases for competing in foreign markets, including competition in quality.

The most spectacular twentieth-century demonstration of the power of competition in quality came from the Japanese. Following World War II, Japanese organizations discovered that the West was unwilling to buy their products—Japan had acquired a reputation for making and exporting shoddy goods. The inability to sell became an alarm signal and a stimulus for launching the Japanese quality revolution during the 1950s. Within a few decades, that revolution propelled Japan into a position of world leadership in quality. This quality leadership in turn enabled Japan to become an economic superpower. It was a phenomenon without precedent in industrial history.

The Twenty-First Century and Quality

The cumulative effect of these massive forces has been to "move quality to center stage." Such a massive move logically should have stimulated a corresponding response—a revolution in managing for quality. However, it was difficult for organizations to recognize the need for such a revolution—they lacked the necessary alarm signals. Technological measures of quality did exist on the shop floors, but managerial measures of quality did not exist in the boardrooms. Thus, except for Japan, the needed quality revolution did not start until very late in the twentieth century. To make this revolution effective throughout the world, economies will require many decades—the entire twenty-first century. Thus, while the twentieth century has been the "century of productivity," the twenty-first century will be known as the "century of quality."

The failure of the West to respond promptly to the need for a revolution in quality led to a widespread crisis. The 1980s then witnessed quality initiatives being taken by large numbers of organizations. Most of these initiatives fell far short of their goals. However, a few were stunningly successful and produced the lessons learned and role models that will serve as guides for the West in the decades ahead.

Today all countries can attain superiority in quality. The methods, tools, and know-how exist. A country that is an emerging country today may provide higher quality than one that has been producing it for centuries. Today, and into the foreseeable future, all organizations in all industries must continue to strive for perfection. They need to be in a state of performance excellence.

The Lessons Learned

Organizations that were successful in their quality initiatives made use of numerous strategies. Analysis shows that despite differences among the organizations, there was much commonality. These common strategies included the following:

1. *Customers and quality must have top priority.* Customer satisfaction is the chief operating goal embedded in the vision and strategic plans. This must be written into corporate policies and scorecards.
2. *Leaders must create a performance excellence system.* All organizations that attained superior results did so with a systematic improvement program or a systematic model for change. This model enables organizational breakthroughs to occur.
3. *Leaders must make quality a strategic priority.* The business plan was opened up to include quality goals and balanced scorecards, year after year.
4. *Continually benchmark best practices.* This approach was adopted to set goals based on superior results already achieved by others.
5. *Engage in continuous innovation and process improvement.* The business plan was opened up to include goals for improvement. It was recognized that quality is a moving target; therefore, there is no end to improving processes.
6. *Offer training in managing for quality, the methods and tools.* Training was extended beyond the quality department to all functions and levels, including upper managers.
7. *Create an organizationwide assurance focus.* This focus is on improving and ensuring that all goods, services, processes, and functions in an organization are of high quality.
8. *Project by project, create multifunctional teams.* Multifunctional teams, adopted to give priority to organization results rather than to functional goals, and later extended to include suppliers and customers, are key to creating breakthroughs in current performance. They focus on the "vital few" opportunities for improvement.
9. *Empower employees.* This includes training and empowering the workforce to participate in planning and improvement of the "useful many" opportunities. Motivation was supplied through extending the use of recognition and rewards for responding to the changes demanded by the quality revolution. Measurements were developed to enable upper managers to follow progress toward providing customer satisfaction, meeting competition, improving quality, and so on. Upper managers took charge of managing for quality by recognizing that certain responsibilities were not delegable—they were to be carried out by the upper managers, personally.

10. *Renew your commitment to superior quality.* Each year competition and technology continue to drive to new levels of customer satisfaction and will result in losing your stature in quality. Annually review the progress and performance of your products and services to ensure the goals are at higher levels of performance as related to quality. The first step to becoming the market quality leader requires that leaders understand the means to create a culture of quality. This is the subject of Chapter 2.

References

"Betting to Win on the Baldie Winners." (1993). *Business Week*, October 18.

Juran, J. M. (1995). *Managerial Breakthrough*. McGraw-Hill, New York.

Juran, J. M. (1999). *Juran's Quality Handbook,* 5th ed. McGraw-Hill, New York.

Juran, J. M. and DeFeo, J. A. (2010). *Juran's Quality Handbook, The Complete Guide to Performance Excellence*, 6th ed. McGraw-Hill, New York.

CHAPTER 2

Three Universal Quality Management Methods

This chapter deals with the fundamental concepts that define the subject of managing for quality. It makes critical distinctions between similar but different contemporary programs to improve performance. It identifies the key processes or the *universals* through which quality is managed and integrated into the strategic fabric of an organization. It demonstrates that while managing for quality is a timeless concept, it has undergone frequent revolution in response to the endless procession of changes and crises faced by our societies.

The Concept of Universals

During Dr. Juran's studies of algebra and geometry, he stumbled across two broad ideas that he put to extensive use in later years. One was the concept of *universals*; the other was the distinction between theory and fact.

His study of algebra exposed him for the first time to the use of symbols to create generalized models. He knew that 3 children plus 4 children added up to 7 children, and 3 beans plus 4 beans added up to 7 beans. Now by using a symbol such as x, he could generalize the problem of adding 3 + 4 and state it as a universal:

$$3x + 4x = 7x$$

This universal said that 3 + 4 always equals 7 no matter what x stands for—children, beans, or anything else. To him the concept of universals was a blinding flash of illumination. He soon found out that universals abounded, but they had to be discovered. They had various names—rules, formulas, laws, models, algorithms, patterns. Once discovered, they could be applied to solve many problems.

By 1954 in his text *Managerial Breakthrough*, Dr. Juran outlined the beginnings of the many universals that led to superior business results. The first was the universal of control—the process for preventing adverse change. The second was the universal sequence for breakthrough improvement. The latter went on to become known as Six Sigma today. By 1986, he discovered his third universal. This was the planning for quality, at the strategic level and at product and service design levels. He also came to realize that those three managerial processes—planning, control, and improvement—were interrelated, so he developed the Juran Trilogy diagram, to depict this interrelationship. The Juran Trilogy embodies the core processes that an organization can use to manage for quality. As a corollary, these same core processes constitute an important sector of science in managing for quality. There is growing awareness in our economy that mastery of those universal processes is critical to attaining leadership in quality and superior results.

What Does Managing for Quality Mean?

For many decades, the phrase used to define quality was simply *fitness for use*. It has been generally accepted that if an organization produced goods that were fit for use as viewed by the customer, then those goods were considered of high quality. Throughout most of the twentieth century this definition made sense because it was easy to grasp. Simply put, if customers purchased a good and it worked, they were pleased with the quality of it. To the producers of that product it was easy to produce as long as the producer had a clear understanding of the customer requirements.

Historically, managing for quality was used to "ensure product conformance to requirements." The majority of tasks largely fell on the supply chain, operations, and the quality departments. These functions were viewed as being responsible to produce, inspect, detect, and ensure the product met requirements.

Two developments have led to modify this time-honored definition of management of quality. The first was the realization that the quality of a physical good—its fitness for use—was broader than just its conformance to specifications. Quality was also determined by the design, packaging, order fulfillment, delivery, field service, and all the service that surrounded the physical good. The operations and quality departments could not manage quality alone because they did not own the resources to do so.

The second development was a shift in the economy from production dominated by goods to production heavily concentrated in services and information. As this shift continued into the twenty-first century, we have chosen to use the phrase *fit for purpose* instead of *fit for use* to define the quality of a product that refers to goods, services, and information. Regardless of whether a product is a good, a service, or information, it must be fit for purpose by the customers of that product. The customer is not just the end user but all those whom the product impacts, including the buyer, the user, the supplier, the regulatory agencies, and almost anyone who is affected by the product from concept to disposal. With such an expanding set of customers and their needs, the methods and tools to manage for quality must grow. Missing important customers and their needs could cause a product to not meet its fitness for purpose definition, and therefore it may not be salable, leading to poor financial performance.

Managing for quality can be defined as "a set of universal methods that any organization, whether a business, an agency, a university, or a hospital, can use to attain superior results by designing, continuously improving, and ensuring that all products, services, and processes meet customer and stakeholder needs."

Management of quality is one set of managerial methods that successful organizations have used to ensure their products—goods, services, and information—meet customer requirements. The evolution from conformance to requirements to fitness for purpose will continue as more industries adopt the methods and tools used to manage for quality. Emerging organizations and countries are creating new means to adopt management methods to their unique needs. Today, a full range of industries, including hospitals, insurance organizations, medical laboratories, and financial service organizations, are managing quality to ensure superior performance.

The accelerated adoption of techniques to manage for quality began in the late 1970s when U.S. businesses were badly affected by many Japanese competitors. Japanese manufactured goods were generally viewed by the purchasers of those goods as having higher quality. This led to the definition of *Japanese* or *Toyota quality*. These terms have become synonymous with higher quality that is required to meet the needs of the customers. As consumers or customers had a better choice, it forced some U.S. organizations into bankruptcy and others to compete at a new level of performance. Eventually many American and then later European organizations regained lost markets with higher quality.

One of the first to accomplish that was Motorola. Motorola was affected by Japanese organizations such as NEC, Sony, and others. The road traveled and the improved quality resulted in Motorola becoming the first winner of the U.S. Malcolm Baldrige National Quality Award. Motorola itself evolved the universal quality improvement model and created the Six Sigma model for quality improvement. Since then, U.S. quality improved, and the quality revolution continued into a global revolution. From 1986 to today, this model of quality improvement has become the most valued model for many industries around the globe. Today organizations such as Toyota, Ford, Samsung, GE, Quest Diagnostics, Mayo Clinic, and Telefonica in Spain have become global quality leaders in their industries.

Each of these organizations also has contributed to the methods to manage quality. Each new organization using the Juran Trilogy adapts it to their unique needs. Many organizations use Lean, Six Sigma, and Quality by Design tools to manage quality into the business processes and all parts of the value chain.

Today most global organizations manage quality from the C suite and not from the quality department. Management of quality is the responsibility of the entire leadership hierarchy. Managing quality has become a driving force behind many business strategies. More organizations are stating in their annual reports that "we will be the best in the industry, we will have the highest quality, and provide the highest level of customer delight." If achieved, these goals will enable these organizations to attain financial success, cultural change, and satisfied customers.

As the needs of customers and society have changed, the means for meeting their needs also changed. The methods of managing quality in 1980 may not work for your organization today. What works today may not work tomorrow. Even the universals that continue to deliver superior results may one day need to be modified. One lesson learned was that many organizations that were once quality leaders failed to sustain their successful performance over time. Why did this happen? Did they fail to sustain results because of weak leadership? Was it external forces? Was it a poor execution of their strategies? These questions have haunted many professionals who have had to defend their quality programs. We will try to provide answers to these questions.

We have presented two of the many meanings of the word *quality* as they relate to goods and services. These two are of critical importance to managing for quality:

1. Quality as it relates to how well the features of a service or good meet customer needs and thereby provide them with satisfaction. In this meaning of the word, higher quality usually costs more.
2. Quality as it relates to freedom from failures (deficiencies). In this sense, the meaning of the word is oriented to costs, and "higher quality usually costs less."

By adopting these simple definitions of quality as it relates to goods and services, one can create a systematic approach to manage quality by:

- Creating processes to design goods and services to meet the needs of its stakeholders (external and internal). Every organization must understand what the customers' needs are and then create or design services and goods that meet those needs.
- Creating processes to control quality. Once designed, these services and goods are produced, at which time we must ensure compliance to the design criteria.
- Creating a systematic approach for improving continuously or creating breakthroughs. Services, goods, and the processes that produce them suffer from chronic failures that must be discovered and remedied.
- Creating a function to ensure you continue to do the three things listed above.

By designing quality, controlling it during operations, and then continuously improving on it, any organization can be on its way to becoming a quality organization. The global quality leaders as described above are relentless in their pursuit of ensuring that all their goods and services meet or exceed their customer requirements—but not at all costs. Attaining quality that satisfies customers but not the business stakeholders is not a good business to be in. To be truly a quality organization, the products and services must be produced at costs that are affordable to the producer and its stakeholders. The quality-cost-revenue relationship, however, must be properly understood in making these judgments. Increased feature quality must generate enough revenue to cover the added costs of additional features. But higher quality from lower failures will usually reduce cost and thereby improve financial performance. For organizations that do not generate revenue, feature quality must not cost more than the budget allows, but quality improvement against failures will almost always improve financial health.

By using these two definitions of quality, and by understanding the impact of good or poor quality on an organization's performance, one can create long-term plans to maintain high quality of goods, services, processes, and financial performance. Managing over the long term also requires that the organization set up systems to ensure that the changing needs of its customers are well understood to avoid the failure to sustain performance that plagues even the most successful organizations.

Organizational Effectiveness Programs

Organizational Effectiveness, Lean Six Sigma, Toyota Production System, and Total Quality Management (TQM) are "brand" names for methods, and some may find them synonymous with the universals to manage for quality. As Juran's universals of managing for quality become embedded and used in many new industries, a new brand may be formed. Most of the time these new brands are useful because they help advance the needs to improve performance. Just as the early guilds led to quality standards, society and changing customer needs also require the universals to be adapted. One common problem with the methods to manage quality was found in the service sector. Service organizations always felt that the word *quality* meant product quality. Many services do not see their products as goods. They are services. Therefore they substitute the words *service quality* with *service excellence*. Over time this phrase catches on and we have a new brand. Most of the time, this new brand builds positively on the previous brand. At other times the alterations to the methods result in less positive outcomes and shunning of the brand. This happened to TQM. Total Quality Management was the brand in the 1990s. It was replaced with Six Sigma. Why? The methods of managing for quality were evolving as many organizations were trying to regain competitiveness. The problem with TQM was that it was not measurable or as business-focused as needed.

Over time it lost its luster. However, there were many organizations that improved their performance immensely, and they continue with TQM today. Others move on to the new brand. At the time of this writing Lean Six Sigma and Performance Excellence are in vogue. They too will change over time. In the end it does not matter what you call your processes to manage for quality as long as you do what is needed to attain superior results. The universals live on. No matter the industry, country, or century, meeting and exceeding the needs of your customers will drive your results.

Management of Quality: The Financial and Cultural Benefits

Customer satisfaction and loyalty are only achieved when both dimensions of quality, features and freedom from failures, are managed effectively and efficiently. Both dimensions impact the financial performance as follows.

Features' Effect on Revenue

Revenue can include several types of transactions: (1) money collected from selling a good or service, (2) taxes collected by a government, or (3) donations received by a charity. Whatever the source, the amount of the revenue relates in varying degrees to the ability of the good or service features produced to be valued by the recipient—the customer. In many markets, goods and services with superior features are able to attract greater revenue through some combination of higher market share and premium pricing. Services and products that are not competitive with features often are sold at lower prices.

Failures' Effect on Income

The customer who encounters a deficiency may take action that creates additional cost for the producer, such as file a complaint, return the product, make a claim, or file a lawsuit. The customer also may elect instead (or in addition) to stop buying from the guilty producer as well as to publicize the deficiency and its source. Such actions by multiple customers can do serious damage to a producer's revenue.

Failures' Effect on Cost

Deficient quality creates excess costs associated with poor quality. *Cost of poor quality* (COPQ) is a term that encompasses all the costs that would disappear if there were no failures—no errors, no rework, no field failures, and so on. Juran Institute's research on the cost of poor quality demonstrates that for organizations that are not managing quality aggressively, the level of COPQ is shockingly high.

Calculating the costs of poor quality can be highly valuable for an organization. COPQ shows enterprise leaders just how much poor quality has inflated their costs and consequently reduced their profits. Detailed COPQ calculations provide a road map for rooting out those costs by systematically removing the poor quality that created them.

In the early 1980s, it was common for many business leaders to make a statement that their COPQ was about 20 to 25 percent of sales revenue. This astonishing number was backed up by many independent organizations calculating their own costs. By the year 2003 the COPQ was in the range of 15 to 20 percent for manufacturing organizations, with many achieving even lower levels as the result of systematic programs to reduce it. For service organizations COPQ as a percentage of sales was still a staggering 30 to 35 percent. These numbers included the costs of redoing what had already been done, the excess costs to control poor processes, and the costs to correctly satisfy customers. Failures that occur prior to sale obviously add to a producer's costs. Failures that occur after sale add to customer's costs as well as to producer's costs. In addition, post-sale failures reduce producers' future sales because customers may be less apt to purchase a poor-quality service.

How to Manage for Quality: A Financial Analogy

To manage quality, it is good to begin by establishing a *vision* for the organization, along with policies, goals, and plans to attain that vision. This means that quality goals and policies must be built into the organization's strategic plan. (These matters are treated elsewhere in *Juran's Quality Handbook*, 6th edition, especially in Chapter 7, "Strategic Planning and Deployment: Moving from Good to Great.") Conversion of these goals into results (making quality happen) is then achieved through established managerial processes—sequences of activities that produce the intended results. Managing for quality makes extensive use of three such managerial processes:

- Designing or planning for quality
- Compliance, controlling or ensuring quality
- Improving or creating breakthroughs in quality

These three processes are interrelated and are known as the *Juran Trilogy*. They parallel the processes long used to manage for finance. These financial processes consist of the following:

- *Financial planning.* This process prepares the annual financial and operational budgets. It defines the deeds to be done in the year ahead. It translates those deeds into money—revenue, costs, and profits. It determines the financial benefits of doing all those deeds. The final result establishes the financial goals for the organization and its various divisions and units.
- *Financial control.* This process consists of evaluating actual financial performance, comparing this with the financial goals, and taking action on the difference—the accountant's *variance*. There are numerous subprocesses for financial control: cost control, expense control, risk management, inventory control, and so on.
- *Financial improvement.* This process aims to improve financial results. It takes many forms: cost reduction projects, new facilities, and new-product development to increase sales, mergers, and acquisitions, joint ventures, and so on.

These processes are universal—they provide the basis for financial management, no matter what type of organization it is.

The financial analogy can help leaders realize that they can manage for quality by using the same processes of planning, control, and improvement. Since the concept of the Juran Trilogy is identical to that used in managing for finance, leaders are not required to change their conceptual approach.

Much of their previous training and experience in managing for finance is applicable to managing for quality.

While the conceptual approach does not change, the procedural steps differ. Table 2.1 shows that each of these three managerial processes has its own unique sequence of activities. Each of the three processes is also a universal—it follows an unvarying sequence of steps. Each sequence is applicable in its respective area, no matter what the industry, function, culture, etc.

Table 2.1   Managing for Quality

Quality Planning	Quality Control	Quality Improvement
Establish goals	Determine the control subjects	Prove the need with a business case
Identify who the customers are	Measure actual performance	Establish a project infrastructure
Determine the needs of the customers	Compare actual performance to the targets and goals	Identify the improvement projects
Develop features which respond to customers' needs	Take action on the difference	Establish project teams
Develop processes able to produce the products	Continue to measure and maintain performance	Provide the teams with resources, training, and motivation to: • Diagnose the causes • Stimulate remedies
Establish process controls		
Transfer the plans to the operating forces		Establish controls to hold the gains

The Juran Trilogy Diagram

The three processes of the Juran Trilogy are interrelated. Figure 2.1 shows this interrelationship.

The Juran Trilogy diagram is a graph with time on the horizontal axis and cost of poor quality on the vertical axis. The initial activity is quality planning. The market research function determines who the customers are and what their needs are. The planners or product realization team then develops product features and process designs to respond to those needs. Finally, the planners turn the plans they created over to operations: "Run the process, produce the features, deliver the product to meet the customers' needs."

Chronic and Sporadic

As operations proceed, soon it is evident that the processes that were designed to deliver the good or service are unable to produce 100 per-

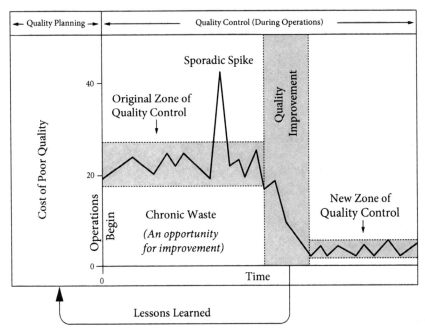

Figure 2.1 Juran Trilogy.

cent quality. Why? Because there are hidden failures or periodic failures that require rework and redoing. Figure 2.1 shows an example where more than 20 percent of the work processes must be redone owing to failures. This waste is considered chronic—it goes on and on until the organization decides to find its root causes. Why do we have this chronic waste? Because it was planned that way. The planners could not account for all unforeseen obstacles in the design process.

Under conventional responsibility patterns, the operating forces are unable to get rid of this planned chronic waste. What they can do is to carry out control—to prevent things from getting worse, as shown in Figure 2.1. The figure shows a sudden sporadic spike that has raised the failure level to more than 40 percent. This spike resulted from some unplanned event such as a power failure, process breakdown, or human error. As a part of the control process, the operating forces converge on the scene and take action to restore the status quo. This is often called *corrective action, troubleshooting, firefighting*, and so on. The end result is to restore the error level back to the planned chronic level of about 20 percent.

The chart also shows that in due course the chronic waste was driven down to a level far below the original level. This gain came from the third process in Juran's Trilogy—improvement. In effect, it was seen that the chronic waste was an opportunity for improvement, and steps were taken to make that improvement.

The Trilogy Diagram and Failures

Juran's Trilogy diagram (Figure 2.1) relates to product and process failures. The vertical scale therefore exhibits units of measure such as cost of poor quality, error rate, percent defective, service call rate, waste, and so on. On this same scale, perfection is at zero, and what goes up is bad. The results of reducing failures are reduction in the cost of poor quality, meeting more delivery promises, reduction of the waste, decrease in customer dissatisfaction, and so on.

References

Itoh, Y. (1978). "Upbringing of Component Suppliers Surrounding Toyota." International Conference on Quality Control, Tokyo.

Juran, J. M. (2004). *Architect of Quality*. McGraw-Hill, New York.

Juran, J. M., and DeFeo, J. A. (2010). *Juran's Quality Handbook: The Complete Guide to Performance Excellence*, 6th ed. McGraw-Hill, New York.

CHAPTER 3

Leadership Role in Creating a Sustainable Culture of Quality

Creating a culture of quality will enable your organization to transform itself from the existing internally focused culture to an externally focused culture. This happens when an organization creates a systematically significant, sustainable, and beneficial change.

The systematic approach that we call the *Juran Transformation Model* can enable any organization to transform itself by knowing what to expect. Transformation usually requires six organizational breakthroughs before a state of performance excellence can be attained.

Culture Defined

Your organization is a society. A *society* is "an enduring and cooperating social group whose members have developed organized patterns of relationships through interaction with each other . . . a group of people engaged in a common purpose."

A society consists of habits and beliefs ingrained over long periods of time. Your workplace is a society, and, as such, it is held together by the shared *beliefs* and *values* that are deeply embedded in the personalities of the society's members. (A workplace whose workforce is segmented into individuals or groups who embody conflicting beliefs and values does not hold together. Various social explosions will eventually occur, including resistances, revolts, mutinies, strikes, resignations, transfers, firings, divestitures, and bankruptcies.)

Society members are rewarded for conforming to their society's beliefs and values—its norms—and they are punished for departing from them. Not only do norms encompass values and beliefs, but also they include enduring systems of relationships, status, customs, rituals, and practices.

Societal norms are so strong and deeply embedded that they lead to customary patterns of social behavior sometimes called *cultural patterns*. In the workplace, one can identify performance-determining cultural patterns such as participative versus authoritarian management styles, casual versus formal dress, conversational styles (Mr./Ms. and Sir/Madam versus first names), and a high trust level that makes it safe to say what you really think versus low trust level/suspiciousness that restricts honest or complete communication and breeds game playing, deceit, and confusion.

What Does Culture Have to Do with Managing an Organization?

To achieve a performance breakthrough, it is desirable—if not necessary—that the organization's norms and cultural patterns support the organization's performance goals. Without this support, performance goals may well be diluted, resisted, indifferently pursued, or simply ignored. For these reasons, the characteristics of your organization's culture are a vital matter that your management needs to understand and be prepared to influence. As we shall see, this is easier said than done; but it *can* be done.

Transforming a Culture

Changing a culture is difficult and usually unsuccessful unless a comprehensive approach exists to achieve and sustain it. The Juran Transformation Model and Roadmap describes five separate and unique types of breakthroughs that must occur in an organization before sustainability is attained. Without these breakthroughs, an organization attains superior results, but the results may not be sustainable for long periods of time. If performance excellence is the state in which an organization attains superior results through the application of the universal quality management methods, then an organization must ensure that these methods are used successfully. The journey from where your organization is to where it wants to go may require a transformational change. This change will result in the ability of the organization to sustain its performance, attain world-class status, and market leadership.

The five breakthroughs are as follows:

1. Leadership and management
2. Organization and structure
3. Current performance
4. Culture
5. Adaptability

The Juran Transformation Model (Figure 3.1) is based on over 60 years of experience and research from Dr. Juran and the Juran Institute. The five breakthroughs, when complete, help produce a state of performance excellence. Each breakthrough addresses a specific organizational subsystem that must change. Each is essential for supporting organizational life; none by itself is sufficient. In effect, the breakthroughs all empower the operational subsystem whose mission is to achieve technological proficiency in producing the goods, services, and information for which customers will pay for or use. There is some overlap and duplication of activities and tasks among the different breakthrough types. This is to be expected because each subsystem is interrelated with all the others, and each is

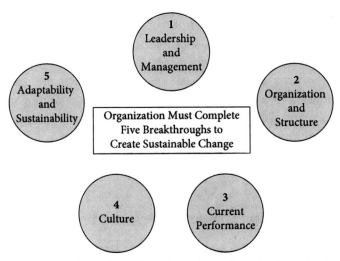

Figure 3.1 Juran Transformation Model. (From Juran Institute, Inc., 2009)

affected by activities in the others. The authors acknowledge that some issues in each type of breakthrough may have already been addressed by the reader's organization—so much the better. If this is the case, if you did not start your organization's performance excellence journey from the beginning, pick up the journey from where your organization presently finds itself. Closing the gaps will likely be part of your organization's next strategic business planning cycle. To close the gaps, design strategic and operational goals and projects to reach those goals and deploy them to all functions and levels.

Breakthrough and Transformational Change

Breakthroughs can occur in an organization at any time, usually as the result of a specific initiative, such as a specific improvement project (e.g., a Six Sigma improvement project, a design of a new service, or the invention of a new technology). These changes can produce sudden explosive bursts of beneficial change for your organization and society. But they may not be enough to cause the culture to change or sustain itself to the changes that occurred. This is so because it may not have happened for the right reason. It was not purposeful. It came about through chance. Change by "chance" is not predictable or sustainable. What an organization needs is predictable change.

Today's organizations operate in a state of perpetual, unpredictable change that requires the people in them to produce continuous adaptive improvements as pressure mounts for new improvements to be made from the outside. These improvements may take months or even years to accomplish because it is the cumulative effect of many coordinated and interrelated organizational plans, policies, and breakthrough projects. Taken together, these diligent efforts gradually transform the organization.

Organizations that do not intend to change usually will when a crisis—or a fear of impending crisis—triggers a need for change within an organization.

Breakthroughs Are Essential to Organizational Vitality

There are four important reasons why an organization cannot survive very long without the medicinal renewing effects of continual breakthrough:

1. The *costs of poor quality* (COPQ) continue to increase if they are not tackled. They are too high. One reason is that organizations are plagued by a continuous onslaught of crises precipitated by mysterious sources of chronic high costs of poorly performing processes. It is not unusual for these costs at times to exceed profit or be a major contributor to losses. In any case, the average overall level is appalling (because it is substantial and *avoidable*), and the toll it takes on the organization can be devastating. COPQ is a major driver of many cost-cutting initiatives, not only because it can be so destructive if left unaddressed but also because savings realized by reducing COPQ directly affect the bottom line. Furthermore, the savings continue, year after year, as long as remedial improvements are irreversible, or controls are placed on reversible improvements.

2. It makes good business sense that *mysterious and chronic causes of waste must be discovered, removed, and prevented from returning*. Breakthrough improvement becomes the preferred initial method of attack because of its ability to uncover and remove specific root causes and to hold the gains—it is designed to do just that. One could describe breakthrough improvement methodology as applying the scientific method to solving performance problems.

3. *Chronic and continuous change.* Another reason why breakthroughs are required for organizational survival is the state of chronic accelerating change found in today's business environment. Unrelenting change has become so powerful and so pervasive that no constituent part of an organization finds itself immune from its effects for long. Because any or all components of an organization can be threatened by changes in the environment, if an organization wishes to survive, it is most likely to be forced into creating basic changes that are powerful enough to bring about accommodation with new conditions. Performance breakthroughs, consisting as they do of several specific types of breakthrough in various organization functions, are a powerful approach that is capable of determining countermeasures sufficiently effective to prevail against the inexorable forces of change. An organization may have to reinvent itself. It may even be driven to reexamine, and perhaps modify, its core products, business, service, or even its customers.

4. *Without continuous improvement, organizations die.* Another reason why breakthroughs are essential for organizational survival is found in

knowledge derived from scientific research into the behavior of organizations. Leaders can learn valuable lessons about how organizations function and how to manage them by examining open systems theory. The Juran Transformation Model is a means by which organizations can stave off their own extinction.

Systems Thinking and Transformational Change

Organizations are like living organisms. They consist of a number of subsystems, each of which performs a vital specialized function that makes specific, unique, and essential contributions to the life of the whole. A given individual subsystem is devoted to its own specific function such as design, production, management, maintenance, sales, procurement, and adaptability. One cannot carry the biological analogy very far because living organisms separate subsystems with physical boundaries and structures (e.g., cell walls, the nervous system, the digestive system, the circulatory system). Boundaries and the structure of subsystems in human organizations, on the other hand, are not physical; they are repetitive events, activities, and transactions. The repetitive patterns of activities are, in effect, the work tasks, procedures, and processes carried out by organizational functions. Open systems theorists call these patterns of activities *roles*. A role consists of one or more recurrent activities out of a total pattern of activities which, in combination, produce the organizational output.

Figure 3.2 shows a model that applies equally to an organization as a whole, to individual subsystems and organizational functions (e.g., departments and workstations within the organization), and to individual organizational members performing tasks in any function or level. All these entities perform three more or less simultaneous roles, acting as supplier, processor, and customer. Acting as a processor, charged with the duty of transforming imported energy, organizations receive raw materials—goods, information, and/or services—from their suppliers, who may be located inside or outside the organization. The processor's job consists of transforming the received things into a new product of some kind—goods, information, or service. In turn, the processor supplies the product to his or her customers who may be located within or outside the organization.

Each of these roles requires more than merely the exchange of things. Each role is linked by mutually understood expectations (i.e., specifications, work orders, and procedures) and feedback as to how well the expectations

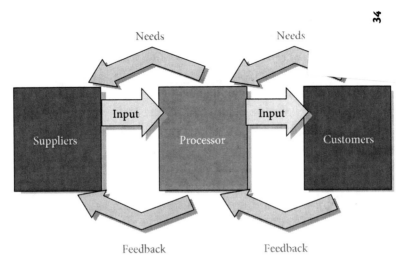

Figure 3.2 The Triple Role. (From Juran Institute, Inc., 2009)

are being met (i.e., complaints, quality reports, praise, and rewards). Note that in the diagram, the processor must communicate (shown by arrows) to the supplier a detailed description of his or her needs and requirements. In addition, the processor provides the supplier with feedback on the extent to which the expectations are being met. This feedback is part of the control loop and helps to ensure consistent adequate performance by the supplier. The customer bears the same responsibilities to his or her processors who, in effect, are also suppliers (not of the raw materials but of the product).

When defects, delays, errors, or excessive costs occur, causes can be found somewhere in the activities performed by suppliers, processors, and customers, in the set of transactions among them, or perhaps in gaps in the communication of needs and feedback. Breakthrough efforts must uncover the precise root causes by deep probing and exploration. If the causes are really elusive, discovering them may require placing the offending repetitive process under a microscope of unprecedented power and precision, as is done in Six Sigma. Performance excellence initiatives require that all functions and levels be involved, at least to some extent, because each function's performance is interrelated to and dependent to some degree on all other functions. Moreover, a change in the behavior of any one function will have some effect on all the others, even though it may not be apparent at the time. This interrelatedness of all functions has practical day-to-day implications for a leader at any level, i.e., the imperative of using "systems thinking" when making decisions, particularly decisions to make changes.

Because an organization is an open system, its life depends on (1) successful transactions with the organization's external environment and (2) proper coordination of the organization's various specialized internal functions and their outputs.

The proper coordination and performance of the various internal functions is dependent on the management processes of planning, controlling, and improving and on human factors such as leadership, organizational structure, and culture. To manage in an open system (such as an organization), management at all levels must think and act in systems terms. Managers must consider the impact of any proposed change not only upon the whole organization but also upon the interrelationships of all the parts. Failure to do so, even when changing seemingly little things, can make some pretty big messes. Leaders need to reason as follows: "If there is to be a change in x, what is required (inputs) from all functions to create this change, and how will x affect each of the other functions, and the total organization as well (ultimate output/results)?" Organizations will not change until the people in them change, regardless of the breakthrough approach. There are three important lessons learned from the experience of the authors:

1. *All organizations need a systematic approach to ensure that change happens.* The problems that appear in one function or step in a process often have their origin upstream from that function or step in the process. People in a given work area cannot necessarily solve the problem in their own work area by themselves—they need to involve others in the problem-solving process. Without systematic involvement of the other functions, suboptimization will occur. Suboptimization results in excess costs and internal customer dissatisfaction—the exact opposite of what is intended.
2. *Change can only be created with active participation of all employees from the top on down and over time.* This includes not only individuals who are the source of a problem but also those affected by the problem and those who will initiate changes to remedy the problem (usually those who are the source of the problem, and perhaps others).

3. *Functional change alone is not sufficient to transform an organization.* Breakthroughs attempted in isolation or within a structure from the whole organization and without systems thinking can easily create more problems than existed at the start of the breakthrough attempt.

Attempts to bring about substantial organizational change such as performance excellence require not only changing the behavior of individuals (as might be attempted by training) but also redefining their roles in the social system. This requires, among other things, changing the expectations that customers have for their processors and changing expectations that processors have for their suppliers. In other words, performance breakthroughs require a capability of organizational design to produce consistent, coordinated behavior to support specific organizational goals. Modifications will likely also be made to other elements that define roles such as job descriptions, job fit, work procedures, control plans, other elements of the quality system, and training. To achieve a breakthrough, it is not sufficient simply to train a few Black Belts in the martial arts as experts and complete a few projects. Although this will probably result in some improvement, it is unlikely to produce long-term culture change and sustainability. The authors believe that too many organizations are settling for simple improvements when they should be striving for breakthroughs.

Attaining a performance excellence state consists of achieving and sustaining beneficial changes. It is noteworthy that having a bright idea for a change does not, by itself, make change actually happen. People must understand why the change is needed and see the impact it will have on them before they can change what they do and perhaps how they do it. Beneficial change is often resisted, sometimes by the very persons who could benefit most from it, especially if they have been successful in doing things the usual way. Leading change can be a perplexing and challenging undertaking. Accordingly, individuals trying to implement change should acquire know-how in how to do it.

Breakthroughs in Leadership and Management

Breakthroughs in leadership occur when managers answer two basic questions:

1. How does management set performance goals for the organization and motivate the people in the organization to reach them and be held accountable?
2. How do managers best use the power of the workforce and other resources in the organization and how should they best manage them?

Issues with leadership are found at *all* levels, not just at the top of an organization. A breakthrough in leadership and management results in an organization characterized by unity of purpose and shared values as well as a system that enables engagement of the workforce.

Each work group knows what its goals are and, specifically, what performance is expected from the team and the individuals. Each individual knows specifically what he or she is to contribute to the overall organizational mission and how his or her performance will be measured. Few erratic or counterproductive behaviors occur. Should such behaviors occur, or should conflict arise, guidelines to behavior and decision making are in place to enable relatively quick and smooth resolution of the problem. There are two major elements to leadership: (1) leaders must decide and clearly communicate where they want their employees to go; and (2) leaders must entice them to follow the path by providing an understanding of why this is a better way. The words *leader* and *manager* do not necessarily refer to different persons. Indeed, most leaders are managers, and managers should be leaders. The distinctions are matters of intent and activities, not players. Leadership can and should be exercised by managers; leaders also need to manage. If leadership consists of influencing others in a positive manner that attracts others, it follows that those at the top of the managerial pyramid (CEOs and C-Suite) can be the most effective leaders because they possess more formal authority than anyone else in an organization. In fact, top managers are usually the most influential leaders. If dramatics change, such as introducing Lean Six Sigma into an organization, the most effective approach by far is for the CEO to lead the charge. Launching Lean Six Sigma is helped immensely if other leaders, such as union presidents, also lead the charge. The same can be said if senior and middle managers, first-line supervisors, and leaders of nonmanagement work crews "follow the leader" and support a performance excellence program by word and actions. Leadership is not

dictatorship because dictators make people afraid of
rect" ways, and perhaps they occasionally provide pu
gasoline, as has happened in Turkmenistan); freeing p...
public spectacles that, together with propaganda, are designed to make
people follow the leader. Dictators do not really get people to want to
behave "correctly" (what the dictator says is correct); the people merely
become afraid not to.

Breakthroughs in Organizational Structure

Creating a breakthrough in organizational structure does the following:

- It designs and puts into place the organization's operational systems (i.e., quality system, orientation of new employees, training, communication processes, and supply chains).
- It designs and puts into practice a formal structure that integrates each function with all the others and sets forth relative authority levels and reporting lines (e.g., organization charts and the means to manage across it).
- It aligns and coordinates the respective interdependent individual functions into a smoothly functioning, integrated organization.

Creating a breakthrough in organizational structure is a response to the basic question, How do I set up organizational structures and processes to reap the most effective and efficient performance toward our goals?

Management structure consists of cross-functional *processes* that are managed by process owners, as well as vertical *functions* that are managed by functional managers. Where both vertical and horizontal responsibility exists, potential conflicts are resolved by matrix mechanisms that require negotiated agreements by the function manager and the cross-functional (horizontal) process owner.

Unity and consistency in the operation of *both* cross-functional processes and vertical functions are essential to creating performance breakthroughs and to continued organizational survival. All members of leadership teams at all levels simply must be in basic agreement as to goals, methods, priorities, and styles. This is especially vital when attempting performance breakthrough improvement projects because the causes of so many performance problems are cross-functional, and the remedies to these problems must be designed and carried out cross-functionally. Consequently, one sees in a Lean or Six Sigma implementation, e.g., quality or executive councils, steering committees, champions (who periodically meet as a group), cross-functional project teams, project team leaders, Black Belts, and Master Black Belts. These roles all involve dealing with change and teamwork issues. There is also a steady trend toward fewer authority or administrative levels and shorter reporting lines.

There are three accepted basic types of organization for managing any function work and one newer, emerging approach. The most traditional and accepted organization types are functional, process, and matrix. They are important design baselines because these organizational structures have been tested and studied extensively and their advantages and disadvantages are well known. The newer, emerging organizational designs are network organizations.

Function-Based Organization

In a function-based organization, departments are established based on specialized expertise. Responsibility and accountability for process and results are usually distributed piecemeal among departments. Many firms are organized around functional departments that have a well-defined management hierarchy. This applies both to the major functions (e.g., human resources, finance, operations, marketing, and product development) and to sections within a functional department. Organizing by function has certain advantages—clear responsibilities and efficiency of activities within a function. A function-based organization typically develops and nurtures talent and fosters expertise and excellence within the functions.

Therefore, a function-based organization offers several long-term benefits. However, this organizational form also creates "walls" between the departments. These walls—sometimes visible, sometimes invisible—often cause serious communication barriers. However, function-based organizations can result in a slow, bureaucratic decision-making apparatus as well as the creation of functional business plans and objectives that may be inconsistent with overall strategic business unit plans and objectives. The outcome can be efficient operations *within* each department but with less than optimal results delivered to external (and internal) customers.

Business Process–Managed Organizations

Many organizations are beginning to experiment with an alternative to the function-based organization in response to today's "make it happen fast" world. Businesses are constantly redrawing their lines, work groups, departments, and divisions, even entire companies, trying to increase productivity, reduce cycle time, enhance revenue, or increase customer satisfaction. Increasingly, organizations are being rotated 90 degrees into processed-based organizations.

In a process organization, reporting responsibilities are associated with a process, and accountability is assigned to a process owner. In a process-based organization, each process is provided with the functionally specialized resources necessary.

This eliminates barriers associated with the traditional function-based organization, making it easier to create cross-functional teams to manage the process on an ongoing basis.

Process-based organizations are usually accountable to the business unit or units that receive the benefits of the process under consideration. Therefore, process-based organizations are usually associated with responsiveness, efficiency, and customer focus.

However, over time, pure process-based organizations run the risk of diluting and diminishing the skill level within the various functions. Furthermore, a lack of process standardization can evolve, which can result in inefficiencies and organizational redundancies. Additionally, such organizations frequently require a matrix-reporting structure, which can result in confusion if the various business units have conflicting objectives. The matrix structure is a hybrid combination of functional and divisional archetypes.

Means of Achieving High Performance

It has been observed that as employees accept more responsibility and have greater motivations, and greater knowledge, they freely participate more toward the interests of the business. They begin to truly act as owners, displaying greater discretionary effort and initiative. Empowered team members have the authority, the capability, and the desire to understand the organization's direction. Consequently, members feel and behave as if they were owners and are willing to accept greater responsibility. They also have greater knowledge, which further enhances their motivation and willingness to accept responsibility.

Enough progress has been made with various empowered organizations that we can now observe some key features of successful efforts. These have come from experiences of various consultants, visits by the authors to other companies, and published books and articles. These key features can help us learn how to design new organizations or redesign old ones to be more effective. The emphasis is on key features, rather than a prescription of how each organization is to operate in detail. This list is not exhaustive, but it is a helpful checklist, useful for a variety of organizations.

Focus on External Customers

The focus is on the external customers, their needs, and the products or services that satisfy those needs.

- The organization has the structure and job designs in place to reduce variation in process and product.
- There are few organizational layers.
- There is a focus on the business and customers.
- Boundaries are set to reduce variances at the source.
- Networks are strong.
- Communications are free flowing and unobstructed.
- Employees understand who the critical customers are, what their needs are, and how to meet customer needs with their own actions. Thus, all actions are based on satisfying the customer. The employees (e.g., operator, technicians, and plant manager) understand that they work for the customer rather than for the plant manager.
- Supplier and customer input is used to manage the business.

In empowered organizations, managers create an environment to make people great, rather than control them. Successful managers are said to "champion" employees and make them feel good about their jobs, their organization, and themselves.

Breakthroughs in Current Performance

Breakthroughs in current performance (or improvement) do the following:

- They significantly improve current levels of results that an organization is attaining. This happens when a systematic project-by-project improvement system discovers root causes of current chronic problems and implements solutions to eliminate them.
- Breakthroughs devise changes to the "guilty" processes and reduce the costs of poorly performing processes.
- They install new systems and controls to prevent the return of these root causes.

A system to attain breakthroughs in current performance addresses the question, How do we reduce or eliminate things that are wrong with our products or processes, and the associated customer dissatisfaction and high costs (waste) that consume the bottom line? A breakthrough improvement program addresses *quality* problems—failures to meet specific important needs of specific customers, internal and external. (Other types of problems are addressed by other types of breakthroughs.) Lean, Six Sigma, Lean Six Sigma, Root Cause Corrective Action, and other programs need to be part of a systematic approach to improve current performance. These methods address a few specific types of things that always go wrong:

- Excessive number of defects
- Undue number of delays
- Unnecessarily long cycle times
- Unwarranted costs of the resulting rework, scrap, late deliveries, dissatisfied customers, replacement of returned goods, loss of customers, and loss of goodwill

Lean and Six Sigma are methods to improve performance. They are project based and require multifunctional teams to improve current levels of performance. Each requires a systematic approach to complete the projects.

Breakthroughs in current levels of performance problems are attained using these methods. The Lean and Six Sigma methods will place your ailing processes under a microscope of unprecedented precision and clarity and make it possible to understand and control the relationships among input variables and desired output variables.

Your organization does have a choice as to what "system" to bring to bear on your problems: a "conventional" weapon system (quality improvement) or a "nuclear" system (Six Sigma). The conventional system is perfectly effective with many problems and much cheaper than the more elaborate and demanding nuclear system. The return on investment is considerable from both approaches, but especially so from Six Sigma if your customers are demanding maximum quality levels.

Breakthroughs in current performance solve problems such as an excessive number of defects, excessive delays, excessively long time cycles, and excessive costs.

Breakthroughs in Culture

The result of completing many improvements creates a habit of improvement in the organization. Each improvement starts to create a quality culture because collectively it does the following:

- It creates a set of new behavior standards and social norms that best supports organizational goals and climate.
- It instills in all functions and levels the values and beliefs that guide organizational behavior and decision making.
- It determines organizational cultural patterns such as style (e.g., informal versus formal, flexible versus rigid, congenial versus hostile, entrepreneurial/risk-taking versus passive/risk adverse, rewarding positive feedback versus punishing negative feedback), extent of internal versus external collaboration, and high energy/morale versus low energy/morale. Performance breakthrough in culture is a response to the basic question, How do I create a social climate that encourages organization members to align together eagerly toward the organization's performance goals?

As employees continue to see their leadership "sticking to it," culture change happens. An organization is not yet at a sustainable level or transformational change. There are still issues that must be addressed, including these:

- Review of the organization's vision, mission, and values
- Orientation of new employees and training practices
- Reward and recognition of policies and practices
- Human resource policies and administration
- Quality and customer satisfaction policies
- Fanatic commitment to customers and their satisfaction
- Commitment to continuous improvement
- Standards and conduct codes, including ethics
- No "sacred cows" regarding people, practices, and core business content
- Community benefit and public relations

How Are Norms Acquired?

New members of a society—a baby born into a family or a new employee hired into the workplace—are carefully taught who is who and what is what. In short, these new members are taught the norms and the cultural patterns of that particular society. In time, they discover that complying with the norms and cultural patterns can be satisfying and rewarding. Resistance or violation of the norms and cultural patterns can be very dissatisfying because it brings on disapproval, condemnation, and possibly punishment. If an individual receives a relatively consistent pattern of rewards and punishments over time, the beliefs and the behaviors being rewarded gradually become a part of that individual's personal set of norms, values, and beliefs. Behaviors that are consistently disapproved or punished will gradually be discarded and not repeated. The individual will have become socialized.

How Are Norms Changed?

Note that socialization can take several years to take hold. This is an important prerequisite for successfully changing an organization's culture that must be understood and anticipated by agents of change, such as senior management. The old patterns must be extinguished and replaced by new ones. This takes time and consistent, persistent effort. These are the realities. Consider what anthropologist Margaret Mead has to say about learning new behaviors and beliefs:

An effective way to encourage the learning of new behaviors and attitudes is by consistent prompt attachment of some form of satisfaction to them. This may take the form of consistent praise, approval, privilege, improved social status, strengthened integration with one's group, or material reward. It is particularly important when the desired change is such that the advantages are slow to materialize—for example, it takes months or even years to appreciate a change in nutrition, or to register the effect of a new way of planting seedlings in the increased yield of an orchard. Here the gap between the new behavior and results, which will not reinforce the behavior until they are fully appreciated, has to be filled in other ways.

She continues:

The learning of new behaviors and attitudes can be achieved by the learner's living through a long series of situations in which the new behavior is made highly satisfying—without exception if possible—and the old not satisfying.

New information psychologically available to an individual, but contrary to his customary behavior, beliefs, and attitudes, may not even be perceived. Even if he is actually forced to recognize its existence, it may be rationalized away, or almost immediately forgotten.

... As an individual's behavior, beliefs, and attitudes are shared with members of his cultural group, it may be necessary to effect a change in the goals or systems of behavior of the whole group before any given individual's behavior will change in some particular respect. This is particularly likely to be so if the need of the individual for group acceptance is very great—either because of his own psychological make-up or because of his position in society.

Implications for achieving breakthroughs in culture require that the leadership team at all levels share, exhibit, and reinforce desired new cultural norms and patterns of behavior—and the norms must be consistent, uninterrupted, and persistent.

Do not expect cultural norms or behavior to change simply because you publish the organization's stated values in official printed material or describe them in speeches or exhortations. Actual cultural norms and patterns may bear no resemblance at all to the values described to the public or proclaimed in exhortations. The same is true of the actual flow of influence compared to the flow shown on the organization chart. (New employees rapidly learn who is really who and what is really what, in contrast to and in spite of the official publicity.)

A forceful leader-manager can, by virtue of his or her personality and commitment, influence the behavior of individual followers in the *short term* with rewards, recognition, and selective exclusion from rewards. The authors know of organizations that, in introducing a Six Sigma or similar effort, have presented messages to their employees along the following lines:

> The organization cannot tell you what to believe, and we are not asking you to believe in our new Six Sigma initiative, although we hope you do. We can, however, expect you to behave in certain ways with respect to it. Therefore, let it be known that you are expected to support it, or at least get out of its way, and not resist. Henceforth, rewards and promotions will go to those who energetically support and participate in the Six Sigma activities. Those who do not support it and participate in it will not be eligible for raises or promotions. They will be left behind, and perhaps even replaced with others who do support it.

This is fairly strong language. Such companies often achieve some results in the short term. However, should a forceful leader depart without causing the new initiative to become embedded in the organization's cultural norms and patterns (to the extent that individual members have taken on these new values and practices as their own), it is not unusual for the new thrust to die out for lack of consistent and persistent reinforcement.

Resistance to Change

Curiously, even with such reinforcement, change—even beneficial change—will often be resisted. The would-be agent of change needs to understand the nature of this resistance and how to prevent or overcome it.

The example of the control chart case drew the conclusion that the main resistance to change is due to the disturbance of the cultural pattern of the shop when a change is proposed or attempted. People who are successful—and therefore comfortable—functioning in the current social or technical system do not want to have their comfortable existence disrupted, especially by an "illegitimate" change.

When a technical or social change is introduced into a group, group members immediately worry that their secure status and comfort level under the new system may be very different (worse) than under the current system. Threatened with the frightening possibility of losing the ability to perform well or losing status, the natural human impulse is to resist the change. Group members have too much at stake in the current system. The new system will require them not only to let go of the current system willingly but also to embrace the uncertain, unpredictable new way of performing. This is a tall order. It is remarkable how profoundly even a tiny departure from cultural norms will upset society members.

What Does Resistance to Change Look Like? Some resistance is intense, dramatic, and even violent. Dr. Juran reminds us of some examples: When fourteenth-century European astronomers postulated a sun-centered universe, this idea flew in the face of the prevailing cultural beliefs in an Earth-centered universe. This belief had been passed down for many generations by their ancestors, religious leaders, grandparents, and parents. (Furthermore, on clear days, one could see with one's own eyes the sun moving around the Earth.) Reaction to the new "preposterous," unacceptable idea was swift and violent. If the sun-centered believers are correct, then the Earth-centered believers are incorrect—an unacceptable, illegitimate, wrong-headed notion. To believe in the new idea required rejecting and tossing out the old. But the old was deeply embedded in the culture. So the "blasphemous" astronomers were burned at the stake.

Another example comes from Dr. Juran: When railroads converted from steam-powered to diesel-powered locomotives in the 1940s, railroad workers in the United States objected. It is unsafe, even immoral, they protested, to trust an entire trainload of people or valuable goods to the lone operator required to drive a diesel. Locomotives had "always" been operated by two people, an engineer who drove and a fireman who stoked the fire. If one *were* incapacitated, the other could take over. But what if the diesel engineer had a heart attack and died? So intense were the resulting strikes that an agreement was finally hammered out to keep the fireman on the job in the diesels! Of course, the railroad workers were really protesting the likely loss of their status and jobs.

Norms Helpful in Achieving a Cultural Transformation   Transforming a culture requires a highly supportive workforce. Certain cultural norms appear to be instrumental in providing the support needed. If these norms are not now part of your culture, some breakthroughs in culture may be required to implant them. Some of the more enabling norms are as follows:

- *A belief that the quality of a product or process is at least of equal importance to, and probably of greater importance than, the mere quantity produced.* This belief results in decisions favoring quality: defective items do not get passed on down the line or out the door; chronic errors and delays are corrected.
- *A fanatical commitment to meeting customer needs.* Everyone knows who his or her customers are (those who receive the results of their work), and how well he or she is doing at meeting those needs. (They ask.) Organization members, if necessary, drop everything and go out of their way to assist customers in need.
- *A fanatical commitment to stretch goals and continuous improvement.* There is always an economic opportunity for improving products or processes. Organizations that practice continuous improvement keep up with, or become better than, competitors.

  Organizations that do not practice continuous improvement fall behind and become irrelevant or worse—go out of business. Six Sigma product design and process improvement is capable, if executed properly, of producing superb economical designs and nearly defect-free processes to produce them, resulting in very satisfied customers and sharply reduced costs. The sales and the savings that follow show up directly on the organization's bottom line.
- *A customer-oriented code of conduct and code of ethics.* This code is published, taught in new employee orientations, and taken into consideration in performance ratings and in distributing rewards. Everyone is expected at all times to behave and make decisions in accordance with the code. The code is enforced, if needed, by managers at all levels. The code applies to everyone, even board members—perhaps especially to them considering their power to influence everyone else.

▲ *A belief that continuous adaptive change is not only good but necessary.* To remain alive, organizations must develop a system for discovering social, governmental, international, or technological trends that could impact the organization. In addition, organizations will need to create and to maintain structures and processes that enable a quick, effective response to these newly discovered trends.

Given the difficulty of predicting trends in the fast-moving contemporary world, it becomes vital for organizations to have such processes and structures in place and operating. If you fail to learn and appropriately adapt to what you learn, your organization can be left behind very suddenly and unexpectedly and end up in the scrap heap. The many rusting, abandoned factories the world over testify to the consequences of not keeping up and consequently being left behind.

Policies and Cultural Norms Policies are guides for managerial action and decision making. Organization manuals typically begin with a statement of the organization's quality policy. This statement rates the relative worth that organization members should place on producing high-quality products, as distinguished from the mere quantity of products produced. (*High-quality products* are goods, services, or information that meets important customer needs at the lowest optimum cost with few, if any, defects, delays, or errors.) High-quality products produce customer satisfaction, sales revenue, repeat demand or sales, and low costs of poor quality (unnecessary waste). Here, in that one sentence, are reasons for attempting quality improvement. Including a value statement in your organization's quality manual reinforces some of the instrumental cultural norms and patterns essential for achieving a quality culture and, ultimately, performance breakthroughs.

Keep in mind that if the value statement, designed to be a guide for decision making, is ignored and not enforced, it becomes worthless, except perhaps as a means of deceiving customers and employees in the short term. You can be sure, however, that customers and employees will soon catch on to the truth and dismiss the quality policy, waving it away as a sham that diminishes the whole organization and degrades management credibility.

Breakthroughs in Adaptability

Creating a breakthrough in adaptability and sustainability requires the following:

- Creation of structures and processes that uncover and predict changes or trends in the environment that are potentially promising or threatening to the organization.
- Creation of processes that evaluate information from the environment and refer it to the appropriate organizational person or function.
- Participation in creating an organizational structure that facilitates rapid adaptive action to exploit the promising trends or avoid the threatening disasters.
- A response to the question, How do I prepare my organization to respond quickly and effectively to unexpected change?

The survival of an organization, like that of all open systems, depends on its ability to detect and react to threats and opportunities that present themselves from within and from outside. To detect potential threats and opportunities, an organization must not only gather data and information about what is happening but also discover the (often) elusive meaning and significance the data hold for the organization. Finally, an organization must take appropriate action to minimize the threats and exploit the opportunities gleaned from the data and information.

To do all this will require appropriate organizational structures, some of which may already exist (an intelligence function, using an adaptive cycle, an Information Quality Council), and a data quality system. The Information Quality Council acts, among other things, as a "voice of the market." *Dates* are defined as "facts" (such as name, address, and age) or "measurements of some physical reality, expressed in numbers and units of measure that enable our organization to make effective decisions by." These measurements are the raw material of *information*, which is defined as "answers to questions" or the "meaning revealed by the data, when analyzed." The typical contemporary organization appears to the authors to be awash in data but bereft of useful information. Even when an organization possesses multiple databases, much doubt exists regarding the quality of the data and, therefore, the organization's ability to tell the truth about the question it is supposed to answer.

Managers dispute the reliability of reports, especially if the messages contained in the data are unfavorable. Department heads question the accuracy of financial statements and sales figures, especially when they bring bad tidings.

Often, multiple databases will convey incongruent or contradictory answers to the same question. This is so because each individual database has been designed to answer questions couched in a unique dialect or based on the unique definitions of terms used by one particular department or function, but not all functions. Data often are stored in isolated unpublicized pockets, out of sight of the very people in other functions who could benefit from them if they knew they existed. Anyone who relies on data for making strategic or operational decisions is rendered almost helpless if the data are not available or are untrustworthy. How can a physician decide on a treatment if X-rays and test results are not available? How can the sales team plan promotions when it does not know how its products are selling compared to the competition? What if these same salespeople knew that the very database that could answer their particular questions already exists but is used for the exclusive benefit of another part of the organization? It is clear that making breakthroughs in adaptability is difficult if one cannot get necessary data and information or if one cannot trust the truthfulness of the information one does get. Some organizations for which up-to-date and trustworthy data are absolutely critical go to great lengths to get useful information. However, in spite of their considerable efforts, many organizations nevertheless remain plagued by chronic data quality problems.

The Route to Adaptability: The Adaptive Cycle and Its Prerequisites

Creating a breakthrough in adaptability creates structures and processes that do the following:

- They detect changes or trends in the internal or external environment that are potentially threatening or promising to the organization.
- They interpret and evaluate the information.
- They refer the distilled information to empowered functions or persons within the organization who take action to ward off threats and exploit opportunities. This is a continuous, perpetual cycle.

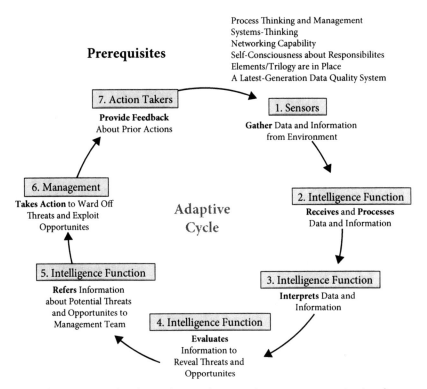

Figure 3.3 Adaptive cycle—to detect and to react to organizational threats and opportunities. (From DeFeo and Barnard, 2004, p. 291)

The cycle might more precisely be conceptualized as a *spiral*, as it goes round and round, never stopping (see Figure 3.3). Several prerequisite actions are needed to set the cycle in motion and create breakthroughs in adaptability. Although each prerequisite is essential, and all are sufficient, perhaps the most crucial is the Information Quality Council and the data quality system. Everything else flows from timely trustworthy data—data that purport to describe truthfully the aspects of reality that are vital to your organization.

Prerequisites for the Adaptive Cycle: Breakthroughs

- Leadership and management
- Organizational structure

- Current performance
- Culture

A Journey Around the Adaptive Cycle

An intelligence function gathers data and information from the internal and external environment. At a minimum, we need to know some of the following basic things.

From the Internal Environment

- Process capability of our measurement and data systems
- Process capability of our key repetitive processes
- Performance of our key repetitive processes (human resources, sales, design, engineering, procurement, logistics, production, storage, transportation, finance, training, yields, defect types and levels, and time cycles)
- Causes of our most important performance problems
- Management instrument panel information: scorecards (performance toward goals)
- Internal costs and costs of poor quality (COPQ)
- Characteristics of our organizational culture (how much it supports or subverts our goals)
- Employee needs
- Employee loyalty

From the External Environment

- Customer needs, now and in the future (what our customers or clients and potential customers or clients want from us or our products)
- Ideal designs of our products (goods, services, and information)
- Customer satisfaction levels
- Customer loyalty levels
- Scientific, technological, social, and governmental trends that can affect us
- Market research and benchmarking findings (us compared to our competition; us compared to best practices)
- Field intelligence findings (how well our products or services perform in use)

You may add to this list other information of vital interest to your particular organization. This list may seem long. It may seem expensive to get all this information. (It can be.) You may be tempted to wave it away as excessive or unnecessary. Nevertheless, if your organization is to survive, there appears to be no alternative but to gather this kind of information, and on a regular, periodic basis. Fortunately, as part of routine control and tracking procedures already in place, your organization probably gathers much of this data and information. Gathering the rest of the information is relatively easy to justify, given the consequences of being unaware of, or deaf or blind to, vital information.

Information about internal affairs is gathered from routine production and quality reports, sales figures, accounts receivable and payable reports, monthly financial reports, shipment figures, inventories, and other standard control and tracking practices. In addition, specially designed surveys—written and interviews—can be used to gain insights into such matters as the state of employee attitudes and needs. A number of these survey instruments are available off the shelf in the marketplace. Formal studies to determine the capability of your measurement systems and your repetitive processes are routinely conducted if you are using Six Sigma in your organization. Even if you do not use Six Sigma, such studies are an integral part of any contemporary quality system. Scorecards are very widely utilized in organizations that carry out annual strategic planning and deployment. The scores provide management with a dashboard, or instrument panel, that indicates warnings of trouble in specific organizational areas. Final reports of operational projects from quality improvement teams, Six Sigma project teams, and other projects undertaken as part of executing the annual strategic business plan are excellent sources of "lessons learned" and ideas for future projects. The tools and techniques for conducting COPQ studies on a continuing basis are widely available. The results of COPQ studies become powerful drivers of new breakthrough projects because they identify specific areas in need of improvement. In sum, materials and tools for gathering information about your organization's internal functioning are widely available and easy to use.

Gathering information about conditions in the external environment is somewhat more complex. Some approaches require considerable know-how and great care. Determining customer needs is an example of an activity that sounds simple but actually requires some know-how to accomplish properly. First, it is proactive. Potential and actual customers are personally approached and asked to describe their needs in terms of benefits they want from a product, services, or information. Many interviewees will describe their needs in terms of a problem to be solved or a product feature. Responses like these must be translated to describe the benefits the interviewee wants, not the problem to be solved or the product feature they would like. Tools and techniques for determining ideal designs of current and future products or services are also available. They require considerable training to acquire the skills, but the payoffs are enormous. The list of such approaches includes Quality by Design, Design for Six Sigma (DFSS), and TRIZ, a technique developed in Russia for projecting future customer needs and product features. Surveys are typically used to get a feel for customer satisfaction. A "feel" may be as close as you can get to knowledge of customer feelings and perceptions. These glimpses can be useful if they reveal distinct patterns of perceptions whereby large proportions of a sample population respond very favorably or very unfavorably to a given issue. Even so, survey results can hardly be considered "data," although they have their uses if suitable cautions are kept in mind. The limitations of survey research methodology cloud the clarity of results from surveys. (What really is the precise difference between a rating of 2 and a rating of 3? A respondent could answer the same question different ways at 8:00 a.m. and at 3:00 p.m., for example. A satisfaction score increase from one month to another could be meaningless if the group of individuals polled in the second month is not the exact same group that was polled in the first month. Even if they were the same individuals, the first objection raised above would still apply to confound the results.)

A more useful approach for gauging customers' "satisfaction," or more precisely, their detailed responses to the products or services they get from you, is the customer loyalty study, which is conducted in person with trained interviewers every six months or so on the same people. The results of this study go way beyond the results from a survey. Results are quantified and visualized. Customers and former customers are asked carefully crafted standard questions about your organization's products and performance. Interviewers probe the responses with follow-up questions and clarifying questions. From the responses, a number of revealing pieces of information are obtained and published graphically. You learn not only the features of your products or services that cause the respondents happiness and unhappiness but also such things as how much improvement of defect X (late deliveries, for example) it would take for former customers to resume doing business with you.

Here is another example. You can graphically depict the amount of sales (volume and revenue) that would result from given amounts of specific types of improvements. You can also learn what specific "bad" things you'd better improve, and the financial consequences of doing so or not doing so. Results from customer loyalty studies are powerful drivers of strategic and tactical planning, and breakthrough improvement activity.

To discover scientific, technological, social, and governmental trends that could affect your organization, you must simply plow through numerous trade publications, journals, news media, websites, and the like, and network as much as possible. Regular searches can be subcontracted so you receive, say, published weekly summaries of information concerning very specific types of issues of vital concern to you. Although there are numerous choices of sources of information concerning trends, there appears to be little choice of whether to acquire such information. The trick is to sort out the useful from the useless information.

A basic product of any intelligence function is to discover how the sales and performance of our organizations' products, services, and sales compare with our competitors and potential competitors. Market research and field intelligence techniques are standard features in most commercial businesses, and books on those topics proliferate.

Completing the adaptive cycle will enable the organization to attain a breakthrough in adaptability that leads to sustainability. Skipping a breakthrough may not indicate a problem in the short term, only in the long term. Consider the economic crisis that hit the global economy in 2008. There were many global organizations that we considered leaders in their markets—when business was good. During the crisis, so many top performers of the past went out of business, were merged with others, or went into bankruptcy only to emerge a different organization. Why did so many organizations have trouble? Our theory was that although these organizations were good at responding to their customer needs, they were not watching societies' needs. This led to a lack of information that, if it had been available, would have provided enough time to "batten down the hatches," to ride the crisis out. To prevent this from happening, creating a high-performing, adaptable organization may lead to better performance when things are not so good.

References

DeFeo, J. A., and W. W. Barnard. (2004). *Juran Institute's Six Sigma Breakthrough and Beyond: Quality Performance Breakthrough Methods.* McGraw-Hill, New York.

Juran Institute, Inc. (2009). *Quality 101: Basic Concepts and Methods for Attaining and Sustaining High Levels of Performance and Quality,* version 4. Juran Institute, Inc., Southbury, Conn.

CHAPTER 4

Aligning Quality Goals with the Strategic Plan

This chapter describes the means by which an organization must align quality goals with its vision, mission, and strategic plan. The strategic planning and deployment process explains how an organization can integrate and align the methods to attain performance excellence. It addresses such important issues as how to align goals with the organization's vision and mission, how to deploy those goals throughout the organization, and how to derive the benefits of strategic planning.

Strategic Planning and Quality: The Benefits

Strategic planning is the systematic approach to defining long-term business goals and planning the means to achieve them. Once an organization has established its long-term goals, effective strategic planning enables it, year by year, to create an annual business plan, which includes the necessary annual goals, resources, and actions needed to move toward those goals.

Many organizations have created a vision to be the best performers by creating and producing high-quality products and services for their customers. By doing so, they have outperformed those who did not. This performance is related not just to the quality of their goods and services, but to the business itself: more sales, fewer costs, and better culture through employee satisfaction and ultimately better market success for its stakeholders.

It is necessary to incorporate these goals into the strategic planning process and into the annual business plans. This will ensure that the new focus becomes part of the plan and does not compete with the well-established priorities for resources. Otherwise, the best-intended desired changes will fail.

Many leaders understand the meaning of strategic planning as it relates to the creation of the strategic plan and the financial goals and targets to be achieved. Often, they do not include the deployment of strategic *quality* goals, subgoals, and annual goals or the assignment of the resources and actions to achieve them. We will try to highlight this difference and use the term *strategic planning and deployment* throughout this chapter. Many organizations have overcome failures of change programs and have achieved long-lasting results through strategic deployment.

Six Sigma, Lean Six Sigma, and in prior years TQM all became pervasive change processes and were natural candidates for inclusion in the strategic plan of many organizations. The integration of these quality and customer-driven methods with strategic planning is important for their success.

Organizations have chosen different terms for this process. Some have used the Japanese term *hoshin kanri*. Others have partially translated the term and called it *hoshin planning*. Still others have used a rough translation of the term and called it *policy deployment*. In an earlier version of the U.S. Malcolm Baldrige National Quality Award, this process was called *strategic quality planning*. Later this award criterion was renamed *strategic planning*.

Whether the upper managers should align quality with the plan is a decision unique to each organization. What is decisive is the importance of integrating major change initiatives or quality programs into the strategic plan. The potential benefits of strategic planning and deployment are clear:

- ▲ The goals become clear—the planning process forces clarification of any vagueness.
- ▲ The planning process then makes the goals achievable.
- ▲ The monitoring process helps to ensure that the goals are reached.
- ▲ Chronic wastes are *scheduled* to be reduced through the improvement process.
- ▲ Creation of new focus on the customers and quality is attained as progress is made.

What Is Strategic Planning and Deployment?

It is a systematic approach to integrating customer-focused, systemwide quality and business excellence methods into the strategic plan of the organization. Strategic planning is the systematic process by which an organization defines its long-term goals with respect to quality and customers, and integrates them—on an equal basis—with financial, human resources, marketing, and research and development goals into one cohesive business plan. The plan is then deployed throughout the entire organization.

As a component of an effective business management system, strategic planning enables an organization to plan and execute strategic organizational breakthroughs. Over the long term, the intended effect of such breakthroughs is to achieve competitive advantage or to attain a status of *quality leadership*.

Strategic planning has evolved during the past decades to become an integral part of many organizational change processes, such as Six Sigma or Operational Excellence (OpEx). It now is part of the foundation that supports the broader system of managing the business of the organization. A simple strategic planning and deployment model is shown in Figure 4.1. This will be used throughout this chapter.

Strategic planning and deployment also is a key element of the U.S. Malcolm Baldrige National Quality Award and the European Foundation

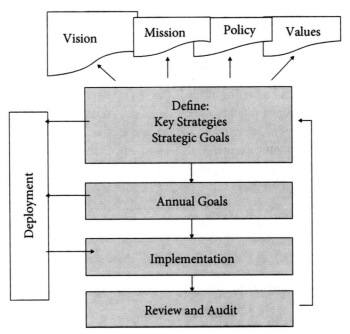

Figure 4.1 Strategic planning model. (From Juran and DeFeo, 2010)

for Quality Management (EFQM) Award, as well as other international and state awards. The criteria for these awards stress that customer-driven quality and operational performance excellence must be key strategic business issues, which need to be an integral part of overall business planning. A critical assessment of the Malcolm Baldrige National Quality Award winners demonstrates that those organizations that won the award outperformed those that did not (Table 4.1).

Table 4.1   Malcolm Baldrige National Quality Award Winner Performance (From *Businessweek*, 1998)

	1988–1996 Investments	Value on 12/1/97	Percent Change
All Recipients	$7,496.54	$33,185.69	342
Standard & Poor's 500	$7,496.54	$18,613.28	148

Data: *National Institute of Standards and Technology*

From 1995 to 2002, quality demonstrated just how profitable it can be. The *Baldrige Index* outperformed the S&P 500 stock index for eight straight years, in certain years beating the S&P by wide margins of 4:1 or 5:1. The index was discontinued in 2004 when Baldrige began to recognize and award small businesses and educational entities along with their normal categories for National Quality Awards. The additions of smaller organizations skewed the "Baldrige Index," yet the results from the original study, when the playing fields were level, speak volumes: quality pays off.

Godfrey (1997) has observed that to be effective, strategic deployment should be used as a tool, a means to an end, not as the goal itself. It should be an endeavor that involves people throughout the organization. It must capture existing activities, not just add to already overflowing plates. It must help senior managers face difficult decisions, set priorities, and not just start new initiatives but eliminate many current activities that add no value.

Quality and Customer Loyalty Goals

These major goals are incorporated and supported by a hierarchy of goals at lower levels: subgoals, projects, etc. Improvement goals are goals aimed at creating a breakthrough in performance of a product, serving process, or people by focusing on the needs of customers, suppliers, and shareholders. The plan incorporates the *voice of the customer* and aligns it to the plan. This alignment enables the goals to be legitimate and balances the financial goals (which are important to shareholders) with those of importance to the customers. It also eliminates the concern that there are two plans, one for finance and one for quality.

A systematic, structured methodology for establishing annual goals and providing resources must include the following:

- *A provision of rewards.* Performance against improvement goals is given substantial weight in the system of merit rating and recognition. A change in the structure that includes rewarding the right behaviors is required.
- *Required and universal participation.* The goals, reports, reviews, etc., are designed to gain participation from within the organization's hierarchy. This participation involves every employee at every level, providing support for the change initiative and helping achieve the desired results.
- *A common language.* Key terms, such as *quality, benchmarking,* and *strategic quality deployment,* acquire standard meanings so that communication becomes more and more precise.
- *Training.* It is common for all employees to undergo training in various concepts, processes, methods, tools, etc. Organizations that have so trained their workforce, in all functions, at all levels, and at the right time, are well poised to outperform organizations in which such training has been confined to the quality department or managers.

Why Strategic Deployment? The Benefits

The first question that often arises in the beginning stages of strategic planning in an organization is, Why do strategic planning in the first place? To answer this question requires a look at the benefits that other organizations have realized from strategic planning. They report that strategic planning:

- Focuses the organization's resources on the activities that are essential to increasing customer satisfaction, lowering costs, and increasing shareholder value (see Table 4.1)

- Creates a planning and implementation system that is responsive, flexible, and disciplined
- Encourages interdepartmental cooperation
- Provides a method to execute breakthroughs year after year
- Empowers leaders, managers, and employees by providing them with the resources to carry out the planned initiatives
- Eliminates unnecessary and wasteful initiatives that are not in the plan
- Eliminates the existence of many potentially conflicting plans—the finance plan, the marketing plan, the technology plan, and the quality plan
- Focuses resources to ensure financial plans are achievable

Why Strategic Deployment? The Risks

Different organizations have tried to implement Total Quality Management systems as well as other change management systems. Some organizations have achieved stunning results; others have been disappointed by their results, often achieving little in the way of bottom-line savings or increased customer satisfaction. Some of these efforts have been classified as failures. One of the primary causes of these disappointments has been the inability to incorporate these "quality programs" into the business plans of the organization.

Other reasons for failure are as follows:

- Strategic planning was assigned to planning departments, not to the upper managers themselves. These planners lacked training in concepts and methods, and were not among the decision makers in the organization. This led to a strategic plan that did not include improvement goals aimed at customer satisfaction, process improvement, etc.
- Individual departments had been pursuing their own departmental goals, failing to integrate them with the overall organizational goals.
- New products or services continued to be designed with failures from prior designs that were carried over into new models, year after year. The new designs were not evaluated or improved and hence were not customer-driven.
- Projects suffered delays and waste due to inadequate participation and ended before positive business results were achieved.

⚠ Improvement goals were assumed to apply only to manufactured goods and manufacturing processes. Customers became irritated by receipt of defective goods; they were also irritated by receiving incorrect invoices and late deliveries. The business processes that produce invoices and deliveries were not subject to modern quality planning and improvement because there were no such goals in the annual plan to do so.

The deficiencies of the past strategic planning processes had their origin in the lack of a systematic, structured approach to integrate programs into one plan. As more organizations became familiar with strategic quality deployment, many adopted its techniques, which treat managing for change on the same organizationwide basis as managing for finance.

Launching Strategic Planning and Deployment

Creating a strategic plan that is quality- and customer-focused requires that leaders become coaches and teachers, personally involved and consistent; eliminate the atmosphere of blame; and make their decisions on the best available data.

Juran (1988) has stated, "You need participation by the people that are going to be impacted, not just in the execution of the plan but in the planning itself. You have to be able to go slow, no surprises, use test sites in order to get an understanding of what are some things that are damaging and correct them."

The Strategic Deployment Process

The strategic deployment process requires that the organization incorporate customer focus into the organization's vision, mission, values, policies, strategies, and long- and short-term goals and projects. Projects are the day-to-day, month-to-month activities that link quality improvement activities, reengineering efforts, and quality planning teams to the organization's business objectives.

The elements needed to establish strategic deployment are generally alike for all organizations. However, each organization's uniqueness will determine the sequence and pace of application and the extent to which additional elements must be provided.

There exists an abundance of jargon used to communicate the strategic deployment process. Depending on the organization, one may use different terms to describe similar concepts. For example, what one organization calls a *vision*, another organization may call a *mission* (see Table 4.2).

Table 4.2   Organizational Vision and Mission (From Juran and DeFeo, 2010)

Selected Definitions	
Mission	What business we are in
Vision	Desired future state of organization
Values	Principles to be observed to meet vision or principle to be served by meeting vision
Policy	How we will operate and our commitment to customers and society

The following definitions of elements of strategic planning are in widespread use and are used in this chapter:

- *Vision.* A desired future state of the organization or enterprise. Imagination and inspiration are important components of a vision. Typically, a vision can be viewed as the ultimate goal of the organization, one that may take five or even ten years to achieve.
- *Mission.* This is the purpose of or the reason for the organization's existence and usually states, for example, what we do and whom we serve.
    - The presence of JetBlue at JFK International is unmatched. Measured by the number of passengers booked, JetBlue carries almost the equivalent of every other airline conducting business at JFK. With its entrenchment in the United States' largest travel market, JetBlue ensures itself profitability even in difficult markets. "Our mission is to bring humanity back to air travel." (JetBlue, 2008)
- *Strategies.* The means to achieve the vision. Strategies are few and define the key success factors, such as price, value, technology, market share, and culture, that the organization must pursue. Strategies are sometimes referred to as *key objectives* or *long-term goals*.
- *Annual goals.* What the organization must achieve over a one- to three-year period; the aim or end to which work effort is directed. Goals are referred to as *long-term* (two to three years) and *short-term* (one to two years). Achievement of goals signals the successful execution of the strategy.

▼ JetBlue aims to preserve the core JetBlue experience of unique, low-cost, high-quality flights while adding optional product offerings for all customers.
▲ *Ethics and values.* What the organization stands for and believes in.
  ▼ For the fourth year in a row, JetBlue was ranked number one in customer service for low-cost carriers by J.D. Power & Associates. It is this exceptional customer service that continues to drive JetBlue and set it apart. Partnerships with Sirius XM, and Direct TV, and improved legroom all make the flight for every customer a more enjoyable experience.
▲ *Policies.* A guide to managerial action. An organization may have policies in a number of areas: quality, environment, safety, human resources, etc. These policies guide day-to-day decision making.
▲ *Initiatives and projects.* These should be multifunctional teams launched to address a deployed goal, and whose successful completion ensures that the strategic goals are achieved. An initiative or project implies assignment of selected individuals to a team, which is given the responsibility, tools, and authority to achieve the specific goal or goals.
  ▼ After six years of planning and three years of construction, JetBlue's Terminal 5 opened at JFK. Terminal 5 offers JetBlue customers their own parking lot and road for improved access to the airliner. It comprises 26 gates, affords the highest in modern amenities and concession offerings, and due to its proximity to the runway is able to be more efficient in its processes. Terminal 5 only advances the company's stake in the New York travel market.
▲ *Deployment plan.* To turn a vision into action, the vision must be broken apart and translated into successively smaller and more specific parts—key strategies, strategic goals, etc.—to the level of projects and even departmental actions. The detailed plan for decomposition and distribution throughout the organization is called the *deployment plan*. It includes the assignment of roles and responsibilities, and the identification of resources needed to implement and achieve the project goals (Figure 4.2).
▲ *Scorecards and key performance indicators.* Measurements that are visible throughout the organization for evaluating the degree to which the strategic plan is being achieved.
  ▼ By the end of 2008, JetBlue was the seventh-largest passenger carrier in the United States and conducted 600 flights daily.

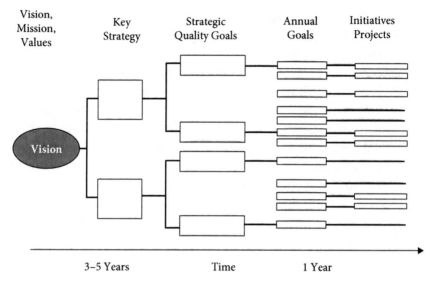

Figure 4.2 Deploying the vision. (From Juran and DeFeo, 2010)

Developing the Elements of Strategic Planning and Deployment

Establish a Vision

Strategic deployment begins with a vision that is customer-focused. In the organizations we know that are successfully making the transition to a more collaborative organization, the key to success lies in developing and living by a common strategic vision. When you agree on an overall direction, you can be flexible about the means to achieve it (Tregoe and Tobia, 1990):

> Really powerful visions are simply told. The Ten Commandments, the Declaration of Independence, a Winston Churchill World War II speech—all present messages that are so simple and direct you can almost touch them. Our corporate strategies should be equally compelling.

A vision should define the benefits a customer, an employee, a shareholder, or society at large can expect from the organization. Here are a few examples:

- Samsung, the world's largest manufacturer of high-quality digital products, is guided by a singular vision: "to lead the digital convergence movement."

  Samsung believes that through technology innovation today, we will find the solutions we need to address the challenges of tomorrow. From technology comes opportunity—for businesses to grow, for citizens in emerging markets to prosper by tapping into the digital economy, and for people to invent new possibilities. It's our aim to develop innovative technologies and efficient processes that create new markets, enrich people's lives, and continue to make Samsung a trusted market leader.
- Sentara Health (based in the mid-Atlantic states), "We have commitment to grow as one of the nation's leading health care organizations by creating innovative systems of care that help people achieve and maintain their best possible state of health."
- Kaiser Permanente (a large U.S.-based health care system): "We are committed to providing our members with quality, cost-effective health care. Our physicians and managers work together to improve care, service, and the overall performance of our organization."

Each of the preceding visions offers a very different view of the direction and character of the organization. Each conveys a general image to customers and employees of where the organization is headed. For the organization, the vision provides, often for the first time in its history, a clear picture of where it is headed and why it is going there.

Good vision statements should also be compelling and shared throughout the organization. It is often a good idea to make the vision a stretch for the organization but possible to be achieved within three to five years, and to state a measurable achievement (e.g., being the best). In creating the vision, organizations should take into account their customers, the markets in which they want to compete, the environment within which the organizations operate, and the current state of the organizations' culture.

Vision statements, by themselves, are little more than words. Publication of such a statement does not inform the members of an organization what they should do differently from what they have done in the past. The strategic deployment process and the strategic plan become the basis for making the vision a reality. The words of the vision are just a reminder of what the organization is pursuing. The vision must be carried out through deeds and actions.

These are some common pitfalls when forming a vision:

- Focusing the vision exclusively on shareholders as customers
- Thinking that once a strategic plan is written, it will be carried out with no further work
- Failing to explain the vision as a benefit to customers, employees, suppliers, and other stakeholders
- Creating a vision that is either too easy or too difficult to achieve
- Failing to consider the effects that the rapid changes taking place in the global economy will have 3 to 5 years into the future
- Failing to involve key employees at all levels in creating the vision
- Failing to benchmark competitors or to consider all possible sources of information on future needs, internal capabilities, and external trends

Agree on Your Mission

Most organizations have a mission statement. A mission statement is designed to address the question, What businesses are we in? A mission is often confused with a vision and even published as one. A mission statement should clarify the organization's purpose or reason for existence. It helps clarify what the organization is.

The following are some examples:

- *Samsung.* Everything we do at Samsung is guided by our mission: to be the best "digital-e-company."
- *Amazon.com.* Our vision is to be earth's most customer-centric company, to build a place where people can come find and discover anything they might want to buy online.
- *Dell.* To be the most successful computer company in the world and deliver the best customer experience in the markets they share.
- *eBay.* To pioneer communities built on commerce, sustained by trust, and inspired by opportunity. eBay brings together millions of people every day on a local, national, and international basis through an array of websites that focus on commerce, payments, and communications.

- *Facebook.* A social utility that helps people communicate more efficiently with their friends, family members, and coworkers. The company develops technologies that facilitate the sharing of information through the social graph, the digital mapping of people's real-world social connections. Anyone can sign up for Facebook and interact with the people they know in a trusted environment.
- *Google.* Organize the world's information and make it universally accessible and useful.
- *Ritz-Carlton Hotel.* A place where the genuine care and comfort of our guests is our highest mission.
- *Sentara Health.* We will focus, plan, and act on our commitments to our community mission, to our customers, and to the highest quality standards of health care to achieve our vision for the future.

In the Sentara example, the references to leadership and the future may lead the reader to confuse this mission statement (what business we are in) with a vision statement (what we aim to become). Only the organization itself can decide whether these words belong in its mission statement. It is in debating such points that an organization comes to a consensus on its vision and mission.

Together, a vision and a mission provide a common agreed-upon direction for the entire organization. This direction can be used as a basis for daily decision making.

To determine what the key strategies should be, one may need to assess five areas of the organization and obtain the necessary data on:

- Customer loyalty and customer satisfaction
- Costs related to poor quality or products, services, and processes
- Organization culture and employee satisfaction
- Internal business processes (including suppliers)
- Competitive benchmarking

Each of these areas, when assessed, can form the basis for a balanced business scorecard (see "The Scorecard" later in this chapter). Data must be analyzed to discover specific strengths, weaknesses, opportunities, and threats as they relate to customers, quality, and costs. Once complete, the key strategies can be created or modified to reflect measurable and observable long-term goals.

Develop Annual Goals

An organization sets specific, measurable strategic goals that must be achieved for the broad strategy to be a success. These quantitative goals will guide the organization's efforts toward achieving each strategy. As used here, a *goal* is an aimed-at target. A goal must be specific. It must be quantifiable (measurable) and is to be met within a specific period. At first, an organization may not know how specific the goal should be. Over time, the measurement systems will improve, and the goal setting will become more specific and more measurable.

Despite the uniqueness of specific industries and organizations, certain goals are widely applicable. There are seven areas that are minimally required to ensure that the proper goals are established:

1. *Product performance.* Goals in this area relate to product features that determine response to customer needs, for example, promptness of service, fuel consumption, mean time between failures, and courteousness. These product features directly influence product salability and affect revenues.
2. *Competitive performance.* This has always been a goal in market-based economies, but seldom a part of the business plan. The trend to make competitive performance a long-term business goal is recent but irreversible. It differs from other goals in that it sets the target relative to the competition, which, in a global economy, is a rapidly moving target. For example, all our products will be considered the "best in class" within one year of introduction as compared to products of the top five competitors.
3. *Business improvement.* Goals in this area may be aimed at improving product deficiencies or process failures, or reducing the cost of poor-quality waste in the system. Improvement goals are deployed through a formal structure of quality improvement projects with assignment of associated responsibilities. Collectively, these projects focus on reducing deficiencies in the organization, thereby leading to improved performance.
4. *Cost of poor quality.* Goals related to quality improvement usually include a goal of reducing the costs due to poor quality or waste in the processes. These costs are not known with precision, though they are estimated to be very high. Nevertheless, it is feasible, through esti-

mates, to bring this goal into the business plan and to deploy it successfully to lower levels. A typical goal is to reduce the cost of poor quality by 50 percent each year for three years.

5. *Performance of business processes.* Goals in this area have only recently entered the strategic business plan. These goals relate to the performance of major processes that are multifunctional in nature, for example, new product development, supply-chain management, and information technology, and subprocesses, such as accounts receivable and purchasing. For such macroprocesses, a special problem is to decide who should have the responsibility for meeting the goal. We discuss this later under "Deployment to Whom?"

6. *Customer satisfaction.* Setting specific goals for customer satisfaction helps keep the organization focused on the customer. Clearly, deployment of these goals requires a good deal of sound data on the current level of satisfaction/dissatisfaction and what factors will contribute to increasing satisfaction and removing dissatisfaction. If the customers' most important needs are known, the organization's strategies can be altered to meet those needs most effectively.

7. *Customer loyalty and retention.* Beyond direct measurement of customer satisfaction, it is even more useful to understand the concept of customer loyalty. Customer loyalty is a measure of customer purchasing behavior between customer and supplier. A customer who needs a product offered by supplier A and who buys solely from that supplier is said to display a loyalty with respect to A of 100 percent. A study of loyalty opens the organization to a better understanding of product salability from the customer's viewpoint and provides the incentive to determine how to better satisfy customer needs. The organization can benchmark to discover the competition's performance, and then set goals to exceed that performance (see Table 4.3).

The goals selected for the annual business plan are chosen from a list of nominations made by all levels of the hierarchy. Only a few of these nominations will survive the screening process and end up as part of the organizationwide business plan. Other nominations may instead enter the business plans at lower levels in the organization. Many nominations will be deferred because they fail to attract the necessary priority and, therefore, will get no organization resources.

Table 4.3 Quality Goals in the Business Plan
(From Juran and DeFeo, 2010)

Product performance (customer focus): This relates to performance features which determine response to customer needs such as promptness of service, fuel consumption, MTBF, and courtesy. (Product includes goods and services.)
Competitive performance: Meeting or exceeding competitive performance has always been a goal. What is new is putting it into the business plan.
Performance improvement: This is a new goal. It is mandated by the fact that the rate of quality improvement decides who will be the quality leader of the future.
Reducing the cost of poor quality: The goal here relates to being competitive as to costs. The measures of cost of poor quality must be based on estimates.
Performance of business processes: This relates to the performance of major multifunctional processes such as billing, purchasing, and launching new products.

Upper managers should become an important source of nominations for strategic goals, since they receive important inputs from sources such as membership on the executive council, contacts with customers, periodic reviews of business performance, contacts with upper managers in other organizations, shareholders, and employee complaints.

Goals that affect product salability and revenue generation should be based primarily on meeting or exceeding marketplace quality. Some of these goals relate to projects that have a long lead time, for example, a new product development involving a cycle time of several years, computerizing a major business process, or a large construction project that will not be commissioned for several years. In such cases, the goal should be set so as to meet the competition estimated to be prevailing when these projects are completed, thereby "leapfrogging" the competition.

In industries that are natural monopolies (e.g., certain utilities), the organizations often are able to make comparisons through use of industry data banks. In some organizations there is internal competition as well—the performances of regional branches are compared with one another.

Some internal departments may also be internal monopolies. However, most internal monopolies have potential competitors—outside suppliers who offer the same services. The performance of the internal supplier can be compared with the proposals offered by an outside supplier.

A third and widely used basis for setting goals has been historical performance. For some products and processes, the historical basis is an aid to needed stability. For other cases, notably those involving high chronic costs of poor quality, the historical basis has done a lot of damage by helping to perpetuate a chronically wasteful performance. During the goal-setting process, upper managers should be on the alert for such misuse of the historical data. Goals for chronically high cost of poor quality should be based on planned breakthroughs using the breakthrough improvement process described in *Juran's Quality Handbook*, 6th ed., Chapter 6, "Quality Improvement: Creating Breakthroughs in Performance."

The Role of Leadership

A fundamental step in the establishment of any strategic plan is the participation of upper management acting as an executive council. Membership typically consists of the key executives. Top-level management must come together as a team to determine and agree upon the strategic direction of the organization. The council is formed to oversee and coordinate all strategic activities aimed at achieving the strategic plan. The council is responsible for executing the strategic business plan and monitoring the key performance indicators. At the highest level of the organization, an executive council should meet monthly or quarterly.

The executives are responsible for ensuring that all business units have a similar council at the subordinate levels of the organization. In such cases, the councils are interlocked; that is, members of upper-level councils serve as chairpersons for lower-level councils (see Figure 4.3).

If a council is not in place, the organization should create one. In a global organization, processes are too complex to be managed functionally. A council ensures a multifunctional team working together to maximize process efficiency and effectiveness. Although this may sound easy, in practice it is not. The senior management team members may not want to give up the monopolies they have enjoyed in the past. For instance, the manager of sales and marketing is accustomed to defining customer needs, the manager of engineering is accustomed to sole responsibility for creating products, and the manager of manufacturing has enjoyed free rein in producing products. In the short run, these managers may not easily give up their monopolies to become team players.

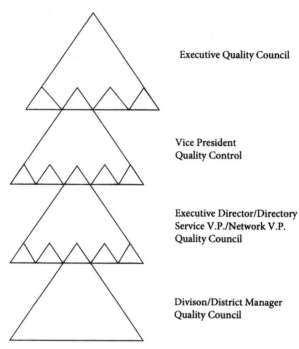

Figure 4.3 How quality councils are linked together. (From Juran, 1994)

Subdivide and Deploy Goals

The deployment of goals is the conversion of goals into operational plans and projects. *Deployment* as used here means subdividing the goals and allocating the subgoals to lower levels. This conversion requires careful attention to such details as the actions needed to meet these goals, who is to take these actions, the resources needed, and the planned timetables and milestones. Successful deployment requires establishment of an infrastructure for managing the plan. Goals are deployed to multifunctional teams, functions, and individuals (see Figure 4.4).

Once the strategic goals have been agreed upon, they must be subdivided and communicated to lower levels. The deployment process also includes dividing up broad goals into manageable pieces (short-term goals or projects). Here are some examples:

- An airline's goal of attaining 99 percent on-time arrivals may require specific short-term (8- to 12-month) initiatives to deal with such matters as these:

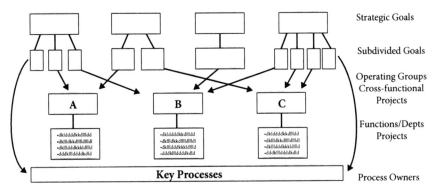

Figure 4.4 Deployment of strategic goals. (From Juran and DeFeo, 2010)

▼ The policy of delaying departures to accommodate delayed connecting flights
▼ The decision making of gate agents at departure gates
▼ The availability of equipment to clean the plane
▼ The need for revisions in departmental procedures to clean the plane
▼ The state of employee behavior and awareness
▲ A hospital's goal of improving the health status of the communities served may require initiatives that:
  ▼ Reduce incidence of preventable disease and illness
  ▼ Improve patient access to care
  ▼ Improve the management of chronic disease conditions
▲ Develop new services and programs in response to community needs

Such deployment accomplishes some essential purposes:

▲ The subdivision continues until it identifies specific deeds to be done
▲ The allocation continues until it assigns specific responsibility for doing the specific deeds

Those who are assigned responsibility respond by determining the resources needed and communicating this to higher levels. Many times, the council must define specific projects, complete with team charters and team members, to ensure goals are met (see Figure 4.5). (For more on the improvement process, see *Juran's Quality Handbook*, 6th ed., Chapter 5, "Quality Improvement: Creating Breakthroughs in Performance.")

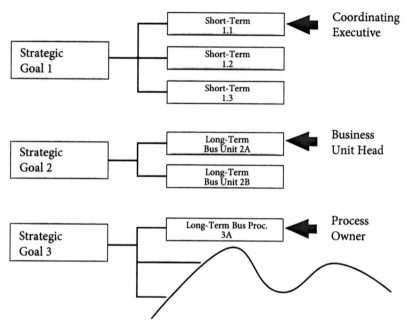

Figure 4.5 Subgoals. (From Juran and DeFeo, 2010)

Deployment to Whom?

The deployment process starts by identifying the needs of the organization and the upper managers. Those needs determine what deeds are required. The deployment process leads to an optimum set of goals through consideration of the resources required. The specific projects to be carried out address the subdivided goals. For example, in the early 1980s, the goal of having the newly designed Ford Taurus/Sable become "best in class" was divided into more than 400 specific subgoals, each related to a specific product feature. The total planning effort was enormous and required more than 1500 project teams.

To some degree, deployment can follow hierarchical lines, such as corporate to division and division to function. However, this simple arrangement fails when goals relate to cross-functional business processes and problems that affect customers.

Major activities of organizations are carried out by the use of interconnecting networks of business processes. Each business process is a multifunctional system consisting of a series of sequential operations. Since it is multifunctional, the process has no single "owner"; hence, there is no obvious answer to the question, Deployment to whom? Deployment is thus made to multifunctional teams. At the conclusion of the team project, an owner is identified. The owner (who may be more than one person) then monitors and maintains this business process. (See *Juran's Quality Handbook*, 6th ed., Chapter 8, "Business Process Management: Creating an Adaptable Organization.")

A Useful Tool for Deployment

The tree diagram is a graphical tool that aids in the deployment process (see Figure 4.5). It displays the hierarchical relationship of the goals, long-term goals, short-term goals, and projects, and indicates where each is assigned in the organization. A tree diagram is useful in visualizing the relationship between goals and objectives or teams and goals. It also provides a visual way to determine if all goals are supported.

Measure Progress with KPIs

There are several reasons why measurement of performance is necessary and why there should be an organized approach to it.

- Performance measures indicate the degree of accomplishment of objectives and, therefore, quantify progress toward the attainment of goals.
- Performance measures are needed to monitor the continuous improvement process, which is central to the changes required to become competitive.
- Measures of individual, team, and business unit performance are required for periodic performance reviews by management.

Once goals have been set and broken down into subgoals, key measures (performance indicators) need to be established. A measurement system that clearly monitors performance against plans has the following properties:

- Indicators link strongly to strategic goals and to the vision and mission of the organization.

- ▲ Indicators include customer concerns; that is, the measures focus on the needs and requirements of internal and external customers.
- ▲ A small number of key measures of key processes can be easily obtained on a timely basis for executive decision making.
- ▲ Chronic waste or cost of poor quality is identified.

For example, Poudre Valley Health Systems (PVHS) established measures of their processes early in the implementation of their business plan and were able to monitor and quantify the following:

- ▲ Improve and maintain employee satisfaction to the top 10 percent of vacancy rate in all U.S. organizations.
- ▲ Strengthen overall service area market share by establishing market strategies specific to service area needs. By breaking down service areas to primary/local and total/national market shares, PVHS aims to control 65 percent of their primary market share and 31.8 percent of total market share by 2012.
- ▲ Support facility development by opening a cancer center.
- ▲ Enhance physician relations by initiating a physician engagement survey tool and reaching a goal of 80 percent satisfaction.
- ▲ Strengthen the company's financial position by achieving a financial flexibility unit of 11 and meeting a five-year plan.

The best measures of the implementation of the strategic planning process are simple, quantitative, and graphical. A basic spreadsheet that describes the key measures and how they will be implemented is shown in Figure 4.6. It is simply a method to monitor the measures.

As goals are set and deployed, the means to achieve them at each level must be analyzed to ensure that they satisfy the objective that they support. Then the proposed resource expenditure must be compared with the proposed result and the benefit/cost ratio assessed. Examples of such measures are as follows:

- ▲ Financial results
  - ▼ Gains
  - ▼ Investment
  - ▼ Return on investment
- ▲ People development
  - ▼ Trained
  - ▼ Active on project teams

Annual Quality Goals	Specific Measurements	Frequency	Format	Data Source	Name

Figure 4.6 Measurement of quality goals. (From Juran and DeFeo, 2010)

- Number of projects
  - Undertaken
  - In process
  - Completed
  - Aborted
- New product or service development
  - Number or percentage of successful product launches
  - Return on investment of new product development effort
  - Cost of developing a product versus the cost of the product it replaces
  - Percentage of revenue attributable to new products
  - Percentage of market share gain attributable to products launched during the last 2 years
  - Percentage of on-time product launches
  - Cost of poor quality associated with new product development
  - Number of engineering changes in the first 12 months of introduction
- Supply-chain management
  - Manufacturing lead times—fill rates
  - Inventory turnover
  - Percentage of on-time delivery

- First-pass yield
- Cost of poor quality

The following is an example of measures that one bank used to monitor teller quality:

- Speed
  - Number of customers in the queue
  - Amount of time in the queue (timeliness)
  - Time per transaction
  - Turnaround time for no-wait or mail transactions
- Accuracy
  - Teller differences in adding up the money at the end of the day
  - Amount charged off/amount handled

Once the measurement system is in place, it must be reviewed periodically to ensure that goals are being met.

Reviewing Progress

A formal, efficient review process will increase the probability of reaching the goals. When planning actions, an organization should look at the gaps between measurement of the current state and the target it is seeking. The review process looks at gaps between what has been achieved and the target (see Figure 4.7).

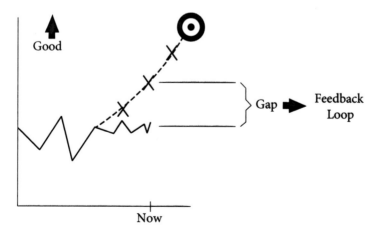

Figure 4.7 Review. (From Juran and DeFeo, 2010)

Frequent measurements of strategic deployment progress displayed in graphical form help identify the gaps in need of attention. Success in closing those gaps depends on a formal feedback loop with clear responsibility and authority for acting on those differences. In addition to the review of results, progress reviews are needed for projects underway to identify potential problems before it is too late to take effective action. Every project should have specific, planned review points, much like those in Figure 4.8.

Organizations today include key performance indicators as discussed in the following sections.

Competitive Quality

These metrics relate to those qualities that influence product salability, for example, promptness of service, responsiveness, courtesy of presale and after-sale service, and order fulfillment accuracy. For automobiles, qualities include top speed, acceleration, braking distance, and safety. For some product features, the needed data must be acquired from customers through negotiation, persuasion, or purchase. For other product features, it is feasible to secure the data through laboratory tests. In still other cases, it is necessary to conduct market research.

Some organizations operate as natural monopolies, for example, regional public utilities. In such cases, the industry association gathers and publishes performance data. In the case of internal monopolies (e.g., payroll preparation, transportation), it is sometimes feasible to secure competitive information from organizations that offer similar services for sale.

Performance on Improvement

This evaluation is important to organizations that go into quality improvement on a project-by-project basis. Due to lack of commonality among the projects, collective evaluation is limited to the summary of such features as:

- *Number of projects.* Undertaken, in-process, completed, aborted.
- *Financial results.* Amounts gained, amounts invested, returns on investment.
- *Persons involved as project team members.* Note that a key measure is the proportion of the organization's management team that is actually

Projects	Project Leaders	Baseline Measurements	Targets	Initial Plan	Review Points				Review Leader
					Resources	Analysis	Plan	Results	

Figure 4.8 Progress review plan. (From Juran and DeFeo, 2010)

involved in improvement projects. Ideally, this proportion should be over 90 percent. In the great majority of organizations, the actual proportion has been less than 10 percent.

Costs of Poor Quality

The *costs of poor quality* are those costs that would disappear if our products and processes were perfect and generated no waste. Those costs are huge. Our research indicates that 15 to 25 percent of all work performed consisted of redoing prior work because products and processes were not perfect.

The costs are not known with precision. In most organizations, the accounting system provides only a small part of the information needed to quantify this cost of poor quality. It takes a great deal of time and effort to extend the accounting system so as to provide full coverage. Most organizations have concluded that such effort is not cost-effective.

The gap can be filled somewhat by estimates that provide upper managers with approximate information as to the total cost of poor quality and the major areas of concentration. These areas of concentration then become the target for quality improvement projects. Thereafter, the completed projects provide fairly precise figures on quality costs before and after the improvements.

Product and Process Failures

Even though the accounting system does not provide for evaluating the cost of poor quality, much evaluation is available through measures of product and process deficiencies, either in natural units of measure or in money equivalents—for example, cost of poor quality per dollar of sales, dollar of cost of sales, hour of work, or unit shipped. Most measures lend themselves to summation at progressively higher levels. This feature enables goals in identical units of measure to be set at multiple levels: corporate, division, and department.

Performance of Business Processes

Despite the wide prevalence and importance of business processes, they have been only recently controlled as to performance. A contributing factor is their multifunctional nature. There is no obvious owner and hence, no clear, sole responsibility for their performance. Responsibility is clear only for the subordinate microprocesses. The system of upper management controls must include control of the macroprocesses. That requires establishing goals in terms of cycle times, deficiencies, etc., and the means for evaluating performances against those goals.

The Scorecard

To enable upper managers to "know the score" relative to achieving strategic quality deployment, it is necessary to design a report package, or scorecard. In effect, the strategic plan dictates the choice of subjects and identifies the measures needed on the upper management scorecard.

The scorecard should consist of several conventional components:

- Key performance indicators (at the highest levels of the organization)
- Quantitative reports on performance, based on data
- Narrative reports on such matters as threats, opportunities, and pertinent events
- Audits conducted (see "Business Audits" later in this chapter)

These conventional components are supplemented as required to deal with the fact that each organization is different. The end result should be a report package that assists upper managers in meeting the quality goals in much the same way as the financial report package assists the upper managers in meeting the financial goals.

Leadership has the ultimate responsibility for designing such a scorecard. In large organizations, design of such a report package requires inputs from the corporate offices and divisional offices alike. At the division level, the inputs should be from multifunctional sources.

The report package should be specially designed to be read at a glance and to permit easy concentration on those exceptional matters that call for attention and action. Reports in tabular form should present the three essentials: goals, actual performances, and variances. Reports in graphical form should, at the least, show the trends of performances against goals. The choice of format should be made only after learning the preferences of the customers, that is, the upper managers.

Managerial reports are usually published monthly or quarterly. The schedule is established to coincide with the meetings schedule of the council or other key reviewing body. The editor of the scorecard is usually the director of quality (quality manager, etc.), who is usually also the secretary of the council.

Scorecards have become an increasing staple in corporations across the globe, so much so that they have moved beyond their initial purpose. Scorecards have now been created not just to document an organization's bottom line but to judge how green an organization actually is. A "Climate Counts" Organization Scorecard rates organizations across different industry sectors in their practices to reduce global warming and create greener business practices. Organizations that are making concerted efforts to alleviate these causes receive higher scores. As in regular scorecards, the information is available to the public, and the opportunity to further a positive public image is at hand. Items include:

- Leading indicators (e.g., quality of purchased components)
- Concurrent indicators (e.g., product test results, process conditions, and service to customers)
- Lagging indicators (e.g., data feedback from customers and returns)
- Data on cost of poor quality

The scorecard should be reviewed formally on a regular schedule. Formality adds legitimacy and status to the reports. Scheduling the reviews adds visibility. The fact that upper managers personally participate in the reviews indicates to the rest of the organization that the reviews are of great importance.

Many organizations have combined their measurements from financial, customer, operational, and human resource areas into *instrument panels* or *balanced business scorecards.*

Business Audits

An essential tool for upper managers is the audit. By *audit* we mean an independent review of performance. *Independent* signifies that the auditors have no direct responsibility for the adequacy of the performance being audited.

The purpose of the audit is to provide independent, unbiased information to the operating managers and others who have a need to know. For certain aspects of performance, those who have a need to know include the upper managers.

To ensure quality, upper management must confirm that:

- The systems are in place and operating properly.
- The desired results are being achieved.

Growing to encompass a broad range of fields, quality audits are now utilized in a plethora of industries, including science. The Royal College of Pathologists implements quality audits on a number of their research reports. The quality audit ensures that individuals and teams are meeting the procedures and standards expected of them and that their work is in line with the mission of the study.

These audits may be based on externally developed criteria, on specific internal objectives, or on some combination of both. Three well-known external sets of criteria to audit organization performance are those of the United States' Malcolm Baldrige National Award for Excellence, the European Foundation Quality Management Award (EFQM), and Japan's Deming Prize. All provide similar criteria for assessing business excellence throughout the entire organization.

Traditionally, quality audits have been used to provide assurance that products conform to specifications and that operations conform to procedures. At upper-management levels, the subject matter of quality audits expands to provide answers to such questions as these:

- Are our policies and goals appropriate to our organization's mission?
- Does our quality provide product satisfaction to our clients?
- Is our quality competitive with the moving target of the marketplace?
- Are we making progress in reducing the cost of poor quality?
- Is the collaboration among our functional departments adequate to ensure optimizing organization performance?
- Are we meeting our responsibilities to society?

Questions such as these are not answered by conventional technological audits. Moreover, the auditors who conduct technological audits seldom have the managerial experience and training needed to conduct business-oriented quality audits. As a consequence, organizations that wish to carry out quality audits oriented to business matters usually do so by using upper managers or outside consultants as auditors.

Juran (1988) has stated:

> One of the things the upper managers should do is maintain an audit of how the processes of managing for achieving the plan is being carried out. Now, when you go into an audit, you have three things to do. One is to identify what are the questions to which we need answers. That's nondelegable; the upper managers have to participate in identifying these questions. Then you have to put together the information that's needed to give the answers to those questions. That can be delegated, and that's most of the work, collecting and analyzing the data. And there's the decisions of what to do in light of those answers. That's nondelegable. That's something the upper managers must participate in.

Audits conducted by executives at the highest levels of the organization where the president personally participates are usually called the *president's audit* (Kondo, 1988). Such audits can have major impacts throughout the organization. The subject matter is so fundamental in nature that the audits reach into every major function. The personal participation of the upper managers simplifies the problem of communicating to the upper levels and increases the likelihood that action will be forthcoming. The very fact that the upper managers participate in person sends a message to the entire organization about the priority placed on quality and the kind of leadership being provided by the upper managers—leading, not cheerleading (Shimoyamada, 1987).

References

Businessweek. (1998). "Malcolm Baldrige National Quality Award Winner Performance." Data: *National Institute of Standards and Technology.* http://www.businessweek.com/

Godfrey, A. B. (1997). "A Short History of Managing Quality in Health Care." In Chip Caldwell, ed., *The Handbook for Managing Change in Health Care.* ASQ Quality Press, Milwaukee, Wis.

JetBlue Airways, Terminal 5, JFK International Airport (2008), http://phx.corporate-ir.net/External.File?item=UGFyZW50SUQ9MzMzODAzfENoaWxkSUQ9MzE2NTMwfFR5cGU9MQ==&t=1.

Juran, J. M. (1988). *Juran on Planning for Quality*. Free Press, New York.

Juran, J. M. (1994). *Managerial Breakthrough*. McGraw-Hill, New York.

Juran, J. M. and DeFeo, J. A. (2010). *Juran's Quality Handbook: The Complete Guide to Performance Excellence*, 6th ed. McGraw-Hill, New York.

Kondo, Y. (1988). "Quality in Japan." In J. M. Juran, ed., *Juran's Quality Control Handbook*, 4th ed. McGraw-Hill, New York. (Kondo provides a detailed discussion of quality audits by Japanese top managers, including the president's audit. See Chapter 35F, "Quality in Japan," under "Internal QC Audit by Top Management.")

Shimoyamada, K. (1987). "The President's Audit: QC Audits at Komatsu." *Quality Progress*, January, pp. 44–49. (Special Audit issue.)

Tregoe, B., and Tobia, P. (1990). "Strategy and the New American Organization." *Industry Week*, August 6.

CHAPTER 5

Product Innovation

New product (goods and services) development is vital to every organization. It is the lifeblood of future sales, performance, and competitiveness. Quality Planning, one of the universals described in previous chapters and used here, is a systematic process for developing new products (both goods and services) and processes that ensure customer needs are met.

The methods to design innovative products have names such as Design for Six Sigma, Design for Lean, Design for World-Class Quality, and Concurrent Engineering, Agile Design for Software. This chapter will focus on the methods and tools that are common to each and sometimes excluded from typical product innovation and development functions. The authors will refer to this as the *Quality by Design method* and tools to enable an organization to develop breakthrough products and services that drive revenue.

Tackling the First Process of the Trilogy: Designing Innovative Products

An organization's ability to satisfy its customers depends on the robustness of the design processes because the goods you sell and the services you offer originate there.

The design process is the first of the three elements of the Juran Trilogy. It is one of the three basic functions by which management ensures the survival of the organization. The design process enables innovation to happen by designing products (goods, services, or information) together with the processes—including controls—to produce the final outputs. When design is completed, the other two elements—control and improvement—kick in to continuously improve upon the design as customer needs and technology change.

89

Juran's universal Quality by Design model has been used since 1986 and provides a structure that can be incorporated into an organization's new product development function; or it can be used independently to be carried out project by project as needed.

The Juran model is especially useful for designing products and redesigning processes simply and economically. The authors have witnessed the design of superb products, processes, and services using this model.

Examples include a prize-winning safety program for a multiple-plant manufacturer; an information system that enables both sales and manufacturing to track the procession of an order throughout the entire order fulfillment process so customers can be informed—on a daily basis—of the exact status of their order; and a redesigned accounts receivable system much faster and more efficient than its predecessor.

The Juran Quality by Design Model

Modern, structured quality design is the methodology used to plan both features that respond to customers' needs and the process to be used to make those features. *Quality by Design* (QbD) refers to the product or service development processes in organizations. Note the dual responsibility of those who plan: to provide the features to meet customer needs and to provide the process to meet operational needs. In times past, the idea that product design stopped at understanding the features that a product should have was the blissful domain of marketers, salespeople, and research and development people. But this new dual responsibility requires that the excitement generated by understanding the features and customer needs be tempered in the fire of operational understanding.

That is, can the processes make the required features without generating waste? To answer this question requires understanding both the current processes' capabilities and customer specifications. If the current processes cannot meet the requirement, modern design must include finding alternative processes that are capable.

The Juran Trilogy points out that the word *quality* incorporates two meanings: first, the presence of features creates customer satisfaction; second, freedom from failures about those features is also needed. In short, failures in features create dissatisfactions.

1. Removing failures is the purpose of quality improvement.
2. Creating features is the purpose of Quality by Design.

Kano, Juran, and others long ago agreed that the absence of failures, that is, no customer dissatisfaction, may not lead us to the belief that satisfaction is thus in hand. We can readily conclude that dissatisfaction goes down as failures are removed. We cannot conclude that satisfaction is therefore going up, because the removal of irritants does not lead to satisfaction—it leads to less dissatisfaction.

It is only the presence of features that creates satisfaction. Satisfaction and dissatisfaction are not co-opposite terms. It is amazing how many organizations fail to grasp this point. Let's take, for example, the typical "bingo card" seen in many hotels. These are replete with "closed-ended" questions. For example, they ask, "How well do you like this on a scale of 1 to 5?" They do not ask, "How much do you like this?" This is the exact opposite of the question "How much do you dislike it?" Therefore, any so-called satisfaction rating that does not allow for open-ended questioning, such as "What should we do that we are not already doing?" or "Is there someone who provides a service we do not offer?" will always fall into a one-sided dimension of quality understanding. What, then, does a composite score of 3.5 for one branch in a chain of hotels really mean compared to another branch scoring 4.0? It means little. Their so-called satisfaction indices are really dissatisfaction indices.

The definition of *fitness for use* takes into account both dimensions of quality—the presence of features and the absence of failures. The sticky points are these: Who gets to decide what *fitness* means? Who decides what *purpose* means? The user decides what *use* means, and the user decides what *fitness* means. Any other answer is bound to lead to argument and misunderstanding. Providers rarely win here. Users, especially society at large, generally always win. For example, take yourself as a consumer. Did you ever use a screwdriver as a pry bar to open a paint can? Of course you did. Did you ever use it to punch holes into a jar lid so your child could watch bugs? Of course you did. Did you ever use it as a chisel to remove some wood, or metal, that was in the way of a job you were doing around the house? Of course you did. Now wait just a moment . . . a screwdriver's intended use is to drive screws!

So the word *use* has two components, *intended* use and *actual* use. When the user utilizes it in the intended way, both the provider and the user are satisfied. Conformance to specification and fitness for purpose match. But what about when the user uses it in the unintended way, as in the screwdriver example? What, then, regarding specifications and fitness?

To delve even deeper, how does the user actually use the product? What need is it meeting for the user? Here we find another juncture: the user can create artful new uses for a product. For example:

> *"2000" Uses for WD-40.* WD-40 was formulated years ago to meet the needs of the U.S. space program. Not many know the origins of the brand name. WD is an acronym for water displacement, and 40 is simply the 40th recipe the company came up with. But as the product moved into the consumer market, all kinds of new uses were uncovered by the users. People claimed it was excellent for removing scuff marks from flooring. They claimed it could easily remove price stickers from lamps, inspection stickers from windshields, and bubble gum from children's hair. The company delighted in all this. But the company didn't release all those clever new uses for public consumption. People also claimed that if they sprayed bait or lures with it, they caught more fish. Those with arthritis swore that a quick spray on a stiff elbow gave them relief. Let's not go too far. What about use where the product obviously cannot work? In Latin there is a word for this: ab-use (or abuse), where the prefix *ab* simply means "not."

Some examples will help: let's go back to the screwdriver. You could argue that using the screwdriver as a pry bar, chisel, or punch is abuse of its original designed purpose. But clearly many manufacturers have provided a product that can withstand this abuse, and so use then falls back into the "intended" column (whether this came as a result of lawsuits or from some other source). Further, a look at commercial aircraft "black boxes" (which are orange, by the way) shows that they clearly survive in circumstances where the aircraft do not survive. Understanding of use in all its forms is what modern design seeks to achieve.

Lastly, modern design and planning, as we see again and again, seeks to create features in response to understanding customer needs. We are referring to *customer-driven* features. The sum of all features is the new product, service, or process.

A different type of product planning in which features meeting no stated need are put out for users to explore is beyond the scope of this chapter. 3M's Post-it Notes and the Internet are examples where we collectively did not voice needs, but we could not imagine life without them, once we embraced their features.

The Quality by Design Problem

The Quality by Design model and its associated methods, tools, and techniques have been developed because in the history of modern society, organizations rather universally have demonstrated a consistent failure to produce the goods and services that unerringly delight their customers. As a customer, everyone has been dismayed time and again when flights are delayed, radioactive contamination spreads, medical treatment is not consistent with best practices, a child's toy fails to function, a new piece of software is not as fast or user-friendly as anticipated, government responds with glacial speed (if at all), or a home washing machine with the latest high-tech gadget delivers at higher cost clothes that are no cleaner than before. These frequent, large quality gaps are really the compound result of a number of four smaller gaps, illustrated in Figure 5.1.

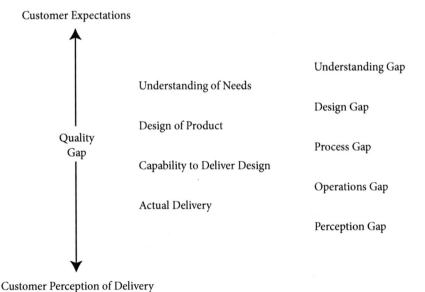

Figure 5.1   The quality gap. (Inspired by Parasuraman et al., 1985, pp. 41–50)

The first component of the quality gap is the *understanding gap*, i.e., lack of understanding of what the customer needs are. Sometimes this gap is wider because the producer simply fails to consider who the customers are and what they need. More often the gap is there because the supplying organization has erroneous confidence in its ability to understand exactly what the customer really needs. The final perception gap in Figure 5.1 also arises from a failure to understand customer needs. Customers do not experience a new suit of clothes or the continuity in service from a local utility simply based on the technical merits of the product. Customers react to how they perceive the good or service provides them with a benefit.

The second constituent of the quality gap is a *design gap*. Even if there were perfect knowledge about customer needs and perceptions, many organizations would fail to create designs for their goods and services that are fully consistent with that understanding. Some of this failure arises from the fact that the people who understand customers and the disciplines they use for understanding customer needs are often systematically isolated from those who actually create the designs. In addition, designers—whether they design sophisticated equipment or delicate human services—often lack the simple tools that would enable them to combine their technical expertise with an understanding of the customer needs to create a truly superior product.

The third gap is the *process gap*. Many splendid designs fail because the process by which the physical product is created or the service is delivered is not capable of conforming to the design consistently time after time. This lack of process capability is one of the most persistent and bedeviling failures in the total quality gap.

The fourth gap is the *operations gap*. The means by which the process is operated and controlled may create additional failures in the delivery of the final good or service.

Quality by Design provides the process, methods, tools, and techniques for closing each of these component gaps and thereby ensuring that any final gap is at a minimum. Table 5.1 summarizes at a high level the basic steps of Quality by Design. The remainder of this section will provide the details and examples for each of these steps.

Juran Quality by Design Model

We look at each of these as we step through the sequence at a high level.

Table 5.1 Quality by Design Steps (From Juran, 2013)

Quality by Design
1. Establish: The project and design goals
2. Define and Identify: The customers
3. Discover: The customer needs
4. Design: The product or service
5. Develop: The process
6. Deliver: The transfer to operations

Step 1—Establish: The Project and Design Goals

A Quality by Design project is the organized work needed to prepare an organization to deliver a new or revised product, service, or process. The following steps or activities are associated with establishing a Quality by Design project:

1. Identify which projects are required to fulfill the organization's sales or revenue generation strategy.
2. Prepare a goal statement for each project.
3. Establish a team to carry out the project.

Identification of Projects

Deciding which projects to undertake is usually the outgrowth of the strategic and business design of an organization. Typically, design for quality projects create new or updated products that are needed to reach specific strategic goals, to meet new or changing customer needs, to fulfill legal or customer mandates, or to take advantage of a new or emerging technology.

Upper management must take the leadership in identifying and supporting the critical Quality by Design projects. Acting as a design council, council, or similar body, management needs to fulfill the following key roles:

1. *Setting design goals.* Marketing, sales, and similar management functions identify market opportunities and client needs currently not being met. By setting these goals, management is beginning the process to create new products, services, or processes to meet these unmet needs.

2. *Nominating and selecting projects.* The management or council selects the appropriate design projects critical to meeting strategic business and customer goals.
3. *Selecting teams.* Once a project has been identified, a team is appointed to see the project through the remaining steps of the design for quality process. A team may be defined by a project manager in the product development function.
4. *Supporting project team.* New technologies and processes are generally required to meet the new design goals. It is up to management to see that each design team is well prepared, trained, and equipped to carry out its goals. The support may include the following:
   a. Provide education and training in design tools.
   b. Provide a trained project leader to help the team work effectively and learn the design for quality process.
   c. Regularly review team progress.
   d. Approve revision of the project goals.
   e. Identify or help with any issues that may hinder the team.
   f. Provide resource expertise in data analysis.
   g. Furnish resources for unusually demanding data collection such as market studies.
   h. Communicate project results.
5. *Monitoring progress.* The council is responsible for keeping the Quality by Design process on track, evaluating progress, and making mid-course corrections to improve the effectiveness of the entire process. Once the council has reviewed the sources for potential projects, it will select one or more for immediate attention. Next, it must prepare a goal statement for the project.

Prepare Goal Statement

Once the council has identified the need for a project, it should prepare a goal statement that incorporates the specific goal(s) of the project. The goal statement is the written charter for the team that describes the intent and purpose of the project. The team goal describes:

- The scope of the project, i.e., the product and markets to be addressed
- The goals of the project, i.e., the results to be achieved (sales targets)

Writing goal statements requires a firm understanding of the driving force behind the project. The goal helps to answer the following questions:

- Why does the organization want to do the project?
- What will the project accomplish once it is implemented?

A goal statement also fosters a consensus among those who either will be affected by the project or will contribute the time and resources necessary to plan and implement the project goal.

Examples include the following:

- The team goal is to deliver to market a new low-energy, fluorocarbon-free refrigerator that is 25 percent less expensive to produce than similar models.
- The team will create accurate control and minimum cost for the inventory of all stores.

While these goal statements describe what will be done, they are still incomplete. They lack the clarity and specificity required of a complete Quality by Design goal statement that incorporates the goal(s) of a project. Well-written and effective goal statements define the scope of the project by including one or more of the following:

Inherent Performance How the final product will perform on one or more dimensions, for example, 24-hour response time, affects the scope of the project.

Comparative Performance How the final product will perform vis-à-vis the competition, e.g., the fastest response time in the metropolitan area, is relevant.

Customer Reaction How will customers rate the product compared with other products available? For example, one organization is rated as having a better on-time delivery service than its closest rival.

Voice of Market Who are or will be the customers or target audience for this product, and what share of the market or market niche will it capture, e.g., to become the "preferred" source by all business travelers within the continental United States?

Performance Failures   How will the product perform with respect to product failure, e.g., failure rate of less than 200 for every 1 million hours of use?

Avoidance of Unnecessary Constraints   It is important to avoid overspecifying the product for the team; e.g., if the product is intended for airline carry-on, specifying the precise dimensions in the goal may be too restrictive. There may be several ways to meet the carry-on market.

Basis for Establishing Quality Goals   In addition to the scope of the project, a goal statement must include the goal(s) of the project. An important consideration in establishing quality goals is the choice of the basis for which the goals are set.

Technology as a Basis   In many organizations, it has been the tradition to establish the quality goals on a technological basis. Most of the goals are published in specifications and procedures that define the quality targets for the supervisory and nonsupervisory levels.

The Market as a Basis   Quality goals that affect product salability should be based primarily on meeting or exceeding market quality. Because the market and the competition undoubtedly will be changing while the design for quality project is underway, goals should be set so as to meet or beat the competition estimated to be prevailing when the project is completed. Some internal suppliers are internal monopolies. Common examples include payroll preparation, facilities maintenance, cafeteria service, and internal transportation. However, most internal monopolies have potential competitors. There are outside suppliers who offer to sell the same service. Thus the performance of the internal supplier can be compared with the proposals offered by an outside supplier.

Benchmarking as a Basis   *Benchmarking* is a recent label for the concept of setting goals based on knowing what has been achieved by others. A common goal is the requirement that the reliability of a new product be at least equal to that of the product it replaces and at least equal to that of the most reliable competing product. Implicit in the use of benchmarking is the concept that the resulting goals are attainable because they have already been attained by others.

History as a Basis   A fourth and widely used basis for setting quality goals has been historical performance; i.e., goals are based on past performance. Sometimes this is tightened up to stimulate improvement. For some products and processes, the historical basis is an aid to needed stability. In other cases, notably those involving chronically high costs of poor quality, the historical basis helps to perpetuate a chronically wasteful performance. During the goal-setting process, the management team should be on the alert for such misuse of the historical basis.

Goals as a Moving Target   It is widely recognized that quality goals must keep shifting to respond to the changes that keep coming over the horizon: new technology, new competition, threats, and opportunities. While organizations that have adopted quality management methods practice this concept, they may not do as well at providing the means to evaluate the impact of those changes and revise the goals accordingly.

Project Goals   Specific goals of the project, i.e., what the project team is to accomplish, are part of an effective goal statement. In getting the job done, the team must mentally start at the finish. The more focused it is on what the end result will look like, the easier it will be to achieve a successful conclusion.

Measurement of the Goal   In addition to stating what will be done and by when, a project goal must show how the team will measure whether it has achieved its stated goals. It is important to spend some time defining how success is measured. Listed below are the four things that can be measured:

1. Quality
2. Quantity
3. Cost
4. Time, speed, agility

New Product Policies   Organizations need to have very clear policy guidelines with respect to quality and product development. Most of these should relate to all new products, but specific policies may relate to individual products, product lines, or groups. Four of the most critical policies are as follows.

1. *Failures in new and carryover designs.* Many organizations have established the clear policy that no new product or component of a product will have a higher rate of failures than the old product or component that it is replacing. In addition, they often require that any carryover design have a certain level of performance; otherwise, it must be replaced with a more reliable design. The minimum carryover reliability may be set by one or more of the following criteria: (1) competitor or benchmark reliability, (2) customer requirements, or (3) a stretch goal beyond benchmark or customer requirements.
2. *Intended versus unintended use.* Should stepladders be designed so that the user can stand on the top step without damage, even though the step is clearly labeled "Do Not Step Here"? Should a hospital design its emergency room to handle volumes of routine, nonemergency patients who show up at its doors? These are policy questions that need to be settled before the project begins. The answers can have a significant impact on the final product, and the answers need to be developed with reference to the organization's strategy and the environment within which its products are used.
3. *Requirement of formal Quality by Design process.* A structured, formal process is required to ensure that the product planners identify their customers and design products and processes that will meet those customer needs with minimum failures. Structured formality is sometimes eschewed as a barrier to creativity. Nothing could be more misguided. Formal Quality by Design identifies the points at which creativity is demanded and then encourages, supports, and enables that creativity. Formal design also ensures that the creativity is focused on the customers and that creative designs ultimately are delivered to the customer free of the destructive influences of failures.
4. *Custody of designs and change control.* Specific provision must be made to ensure that approved designs are documented and accessible. Any changes to designs must be validated, receive appropriate approvals, be documented, and be unerringly incorporated into the product or process. Specific individuals must have the assigned authority, responsibility, and resources to maintain the final designs and administer change control.

Establish a Team

The cross-functional approach to complete a Quality by Design project is effective for several reasons:

- Team involvement promotes sharing of ideas, experiences, and a sense of commitment to being a part of helping "our" organization achieve its goal.
- The diversity of team members brings a more complete working knowledge of the product and processes to be planned. Design of a product requires a thorough understanding of how things get done in many parts of the organization.
- Representation from various departments or functions promotes the acceptance and implementation of the new plan throughout the organization. Products or processes designed with the active participation of the affected areas tend to be technically superior and accepted more readily by those who must implement them.

Step 2—Define and Identify: The Customers

This step may seem unnecessary; of course, the planners and designers know who their customers are—the driver of the automobile, the depositor in the bank account, the patient who takes the medication. But these are not the only customers—not even necessarily the most important customers. Customers comprise an entire cast of characters that needs to be understood fully.

Generally, there are two primary groups of customers: the external customers—those outside the producing organization—and the internal customers—those inside the producing organization.

Types of External Customers

The term *customer* is often used loosely; it can refer to an entire organization, a unit of a larger organization, or a person. There are many types of customers, some obvious, others hidden. Below is a listing of the major categories to help guide complete customer identification.

The Purchaser   This is someone who buys the product for himself or herself or for someone else, e.g., anyone who purchases food for his or her family. The end user/ultimate customer is someone who finally benefits from the product, e.g., the patient who goes to a health care facility for diagnostic testing.

Merchants   Merchants are people who purchase products for resale, wholesalers, distributors, travel agents and brokers, and anyone who handles the product, such as a supermarket employee who places the product on the shelf.

Processors   Processors are organizations and people who use the product or output as an input for producing their own product, e.g., a refinery that receives crude oil and processes it into different products for a variety of customers.

Suppliers   Those who provide input to the process are suppliers, e.g., the manufacturer of the spark plugs for an automobile or the law firm that provides advice on the organization's environmental law matters. Suppliers are also customers. They have information needs with respect to product specification, feedback on failures, predictability of orders, and so on.

Potential Customers   Those not currently using the product but capable of becoming customers are potential customers; e.g., a business traveler renting a car may purchase a similar automobile when the time comes to buy one for personal use.

Hidden Customers   Hidden customers comprise an assortment of different customers who are easily overlooked because they may not come to mind readily. They can exert great influence over the product design: regulators, critics, opinion leaders, testing services, payers, the media, the public at large, those directly or potentially threatened by the product, corporate policymakers, labor unions, and professional associations.

Internal Customers   Everyone inside an organization plays three roles: supplier, processor, and customer. Each individual receives something from someone, does something with it, and passes it to a third individual. Effectiveness in meeting the needs of these internal customers can have a major impact on serving the external customers. Identifying the internal customers will require some analysis because many of these relationships tend to be informal, resulting in a hazy perception of who the customers are and how they will be affected. For example, if an organization decides to introduce just-in-time manufacturing to one of its plants, this will have significant effects on purchasing, shipping, sales, operations, and so on.

Most organizations try to set up a mechanism that will allow seemingly competing functions to negotiate and resolve differences based on the higher goal of satisfying customer needs. This might include conducting weekly meetings of department heads or publishing procedure manuals. However, these mechanisms often do not work because the needs of internal customers are not fully understood, and communication among the functions breaks down. This is why a major goal in the design for quality process is to identify who the internal customers are, discover their needs, and plan how those needs will be satisfied. This is also another reason to have a multifunctional team involved in the planning; these are people who are likely to recognize the vested interests of internal customers.

Identifying Customers

In addition to the general guidance just laid out, it is most often helpful to draw a relatively high-level flow diagram of the processes related to the product being planned. Careful analysis of this flow diagram often will provide new insight, identifying customers that might have been missed and refining understanding of how the customers interact with the process. Figure 5.2 is an example of such a diagram. A review of this diagram reveals that the role of customer is really two different roles—placing the order and using the product. These may or may not be played by the same individuals, but they are two distinct roles, and each needs to be understood in terms of its needs.

Step 3—Discover: Customer Needs

The third step of Quality by Design is to discover the needs of both external customers and internal processors for the product. Some of the key activities required for effective discovery of customer needs include the following:

- ▲ Plan to discover customers' needs.

- Collect a list of customers' needs in their language.
- Analyze and prioritize customers' needs.
- Translate their needs into "our" language.
- Establish units of measurement and sensors.

Our own experience tells us that the needs of human beings are both varied and complex. This can be particularly challenging to a design team because the actions of customers are not always consistent with what they

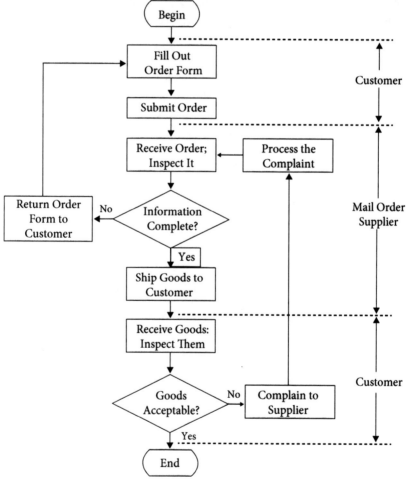

Figure 5.2 Flow diagram and customers. (From Juran, 1999, p. 3.12)

say they want. The challenge for Quality by Design is to identify the most important needs from the full array of those needs expressed or assumed by the customer. Only then can the product delight the customers.

When a product is being designed, there are actually two related but distinct aspects of what is being developed: the technological elements of what the product's features will actually do or how it will function and the human elements of the benefits customers will receive from using the product. The two must be considered together.

Discovering customer needs is a complex task. Experience shows that customers usually do not state, in simple terms, exactly what they want; often they do not even mention some of their most basic needs. Accuracy of bank statements, competence of a physician, reliability of a computer, and grammatical correctness of a publication may be assumed and never stated without probing.

One of the ways customers express their needs is in terms of problems they experience and their expectation that a product will solve their problems. For example, a customer may state, "I cannot always answer my telephone personally, but I do not want callers to be either inconvenienced or disgusted with nonresponsive answering systems." Or the customer may state, "My mother's personal dignity and love of people are very important to me. I want to find an extended care facility that treats her as a person, not a patient." Even when the need is not expressed in such terms, the art and science of discovering needs are to understand exactly the benefit that the customer expects.

When a product's features meet a customer's need, it gives the customer a feeling of satisfaction. If the product fails to deliver the promised feature defect-free, the customer feels dissatisfaction. Even if a product functions the way it has been designed, a competing product, by virtue of superior service or performance, may provide customers with greater satisfaction.

Stated Needs and Real Needs

Customers commonly state their needs as seen from their viewpoint and in their language. Customers may state their needs in terms of the goods or services they wish to buy. However, their real needs are the benefits they believe they will receive.

To illustrate:

Customer wishes to buy:	Benefit customer needs might include:
Fresh pasta	Nourishment and taste
Newest personal computer	Write reports quickly and easily
	Find information on the Web
	Help children learn math
Health insurance	Security against financial disaster
	Access to high-quality health care
	Choice in health care providers
Airline ticket	Transportation, comfort, safety, and convenience

Failure to grasp the difference between stated needs and real needs can undermine the salability of the product in design. Understanding the real needs does not mean that the planners can dismiss the customers' statements and substitute their own superior technical understanding as being the customers' real needs. Understanding the real needs means asking and answering such questions as these:

▲ Why is the customer buying this product?
▲ What service does she or he expect from it?
▲ How will the customer benefit from it?
▲ How does the customer use it?
▲ What has created customer complaints in the past?
▲ Why have customers selected competitors' products over ours?

Perceived Needs

Customers understandably state their needs based on their perceptions. These may differ entirely from the supplier's perceptions of what constitutes product quality. Planners can mislead themselves by considering whether the customers' perceptions are wrong or right rather than focusing on how these perceptions influence customers' buying habits. Although such differences between customers and suppliers are potential troublemakers, they also can be an opportunity. Superior understanding of customer perceptions can lead to competitive advantage.

Cultural Needs

The needs of customers, especially internal customers, go beyond products and processes. They include primary needs for job security, self-respect,

respect of others, continuity of habit patterns, and still other elements of what we broadly call the *cultural values*; these are seldom stated openly. Any proposed change becomes a threat to these important values and hence will be resisted until the nature of the threat is understood.

Needs Traceable to Unintended Use

Many quality failures arise because a customer uses the product in a manner different from that intended by the supplier. This practice takes many forms. Patients visit emergency rooms for nonemergency care. Untrained workers are assigned to processes requiring trained workers. Equipment does not receive specified preventive maintenance.

Factors such as safety may add to the cost, yet they may well result in a reduced overall cost by helping to avoid the higher cost arising from misuse of the product. It is essential to learn the following:

- What will be the actual use (and misuse)?
- What are the associated costs?
- What are the consequences of adhering only to intended use?

Human Safety

Technology places dangerous products into the hands of amateurs who do not always possess the requisite skills to handle them without accidents. It also creates dangerous by-products that threaten human health, safety, and the environment. The extent of all this is so great that much of the effort of product and process design must be directed at reducing these risks to an acceptable level. Numerous laws, criminal and civil, mandate such efforts.

User-Friendly

The amateur status of many users has given rise to the term *user-friendly* to describe the product feature that enables amateurs to make ready use of technological products. For example, the language of published information should be *simple, unambiguous,* and *readily understood.* (Notorious offenders have included legal documents, owners' operating manuals, administrative forms, etc. Widely used forms such as government tax returns should be field-tested on a sample of the very people who will later be faced with filling out such forms.) The language of published information should also be *broadly compatible.* (For example, new releases of software should be "upward-compatible with earlier releases.")

Promptness of Service

Services should be prompt. In our culture, a major element of competition is promptness of service. Interlocking schedules (as in mail delivery or airline travel) are another source of a growing demand for promptness. Still another example is the growing use of just-in-time manufacturing, which requires dependable deliveries of materials to minimize inventories. All such examples demonstrate the need to include the element of promptness in design to meet customer needs.

Customer Needs Related to Failures

In the event of product failure, a new set of customer needs emerges—how to get service restored and how to get compensated for the associated losses and inconvenience. Clearly, the ideal solution to all this is to plan quality so that there will be no failures. At this point, we will look at what customers need when failures do occur.

Warranties

The laws governing sales imply that there are certain warranties given by the supplier. However, in our complex society, it has become necessary to provide specific, written contracts to define just what is covered by the warranty and for how long a time. In addition, it should be clear who has what responsibilities.

Effect of Complaint Handling on Sales

While complaints deal primarily with product dissatisfaction, there is a side effect on salability. Research in this area has pointed out the following: Of the customers who were dissatisfied with products, nearly 70 percent did not complain. The proportions of these who did complain varied according to the type of product involved. The reasons for not complaining were principally (1) the belief that the effort to complain was not worth it, (2) the belief that complaining would do no good, and (3) lack of knowledge about how to complain. More than 40 percent of the complaining customers were unhappy with the responsive action taken by the suppliers. Again, percentages varied according to the type of product.

Future salability is strongly influenced by the action taken on complaints. This strong influence also extends to brand loyalty. Even customers of popular brands of large-ticket items, such as durable goods, financial services, and automobile services, will reduce their intent to buy when they perceive that their complaints are not addressed.

This same research concluded that an organized approach to complaint handling provides a high return on investment. The elements of such an organized approach may include:

- A response center staffed to provide 24-hour access by consumers and/or a toll-free telephone number
- Special training for the employees who answer the telephones
- Active solicitation of complaints to minimize loss of customers in the future

Keeping Customers Informed

Customers are quite sensitive to being victimized by secret actions of a supplier, as the phrase *Let the buyer beware!* implies. When such secrets are later discovered and publicized, the damage to the supplier's quality image can be considerable. In a great many cases, the products are fit for use despite some nonconformances. In other cases, the matter may be debatable. In still other cases, the act of shipment is at least unethical and at worst illegal.

Customers also have a need to be kept informed in many cases involving product failures. There are many situations in which an interruption in service will force customers to wait for an indefinite period until service is restored. Obvious examples are power outages and delays in public transportation. In all such cases, the customers become restive. They are unable to solve the problem—they must leave that to the supplier. Yet they want to be kept informed as to the nature of the problem and especially as to the likely time of solution. Many suppliers are derelict in keeping customers informed and thereby suffer a decline in their quality image. In contrast, some airlines go to great pains to keep their customers informed of the reasons for a delay and of the progress being made in providing a remedy.

Collect a List of Customers' Needs in Their Language

For a list of customers' needs to have significant meaning in the design of a new product, the needs must be stated in terms of benefits sought. Another way of saying this is to capture needs in the customer's voice. By focusing on the benefits sought by the customer rather than on the means of delivering the benefit, designers will gain a better understanding of what the customer needs and how the customer will be using the product. Stating needs in terms of the benefits sought also can reveal opportunities for improved quality that often cannot be seen when concentrating on the features alone.

Analyze and Prioritize Customer Needs

The information actually collected from customers is often too broad, too vague, and too voluminous to be used directly in designing a product. Both specificity and priority are needed to ensure that the design really meets the needs and that time is spent on designing for those needs that are truly the most important. The following activities help provide this precision and focus:

- Organizing, consolidating, and prioritizing the list of needs for both internal and external customers
- Determining the importance of each need for both internal and external customers
- Breaking down each need into precise terms so that a specific design response can be identified
- Translating these needs into the supplying organization's language
- Establishing specific measurements and measurement methods for each need

One of the best design tools to analyze and organize customers' needs is the design for quality spreadsheet.

Quality by Design Spreadsheets

Designing new products can generate large amounts of information that is both useful and necessary, but without a systematic way to approach the organization and analysis of this information, the design team may be overwhelmed by the volume and miss the message it contains.

Although planners have developed various approaches for organizing all this information, the most convenient and basic design tool is the Quality by Design spreadsheet. The spreadsheet is a highly versatile tool that can be adapted to a number of situations. The Quality by Design process makes use of several kinds of spreadsheets, such as:

- Customer needs spreadsheet
- Needs analysis spreadsheet
- Product or service design spreadsheet
- Process design spreadsheet
- Process control spreadsheet

Besides recording information, these tools are particularly useful in analyzing relationships among the data that have been collected and in facilitating the stepwise conversion of customer needs into features and then features into process characteristics and plans. This conversion is illustrated in Figure 5.3. Analysis of customers and their needs provides the basis for designing the product. The summary of that design feeds the process design, which feeds the control spreadsheet.

Translate Their Needs into "Our" Language

The precise customer needs that have been identified may be stated in any of several languages, including:

- The customer's language
- The supplier's ("our") language
- A common language

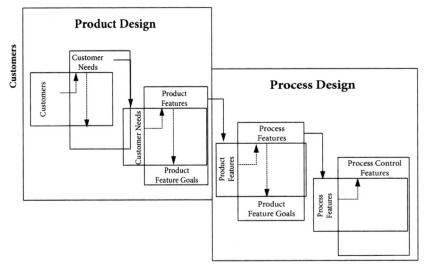

Figure 5.3 Sequence of activities. (From Juran Institute, Inc., 2013)

An old aphorism claims that the British and Americans are separated by a common language. The appearance of a common language or dialect can be an invitation to trouble because both parties believe that they understand each other and expect to be understood. Failure to communicate because of the unrecognized differences can build additional misunderstanding that only compounds the difficulty. It is imperative, therefore, for planners to take extraordinary steps to ensure that they properly understand customer needs by systematically translating them. The need to translate applies to both internal and external customers. Various organizational functions employ local dialects that are often not understood by other functions.

Vague terminology constitutes one special case for translation that can arise even (and often especially) between customers and suppliers who believe they are speaking the same dialect. Identical words have multiple meanings. Descriptive words do not describe with technological precision.

Translating and Measuring Customer Needs

The customer need for performance illustrates how high-level needs break down into myriad detailed needs. Performance included all the following detailed, precise needs:

Product Design Spreadsheet All the information on the translation and measurement of a customer need must be recorded and organized. Experience recommends placing these data so that they will be close at hand during product design. The example in Figure 5.4 shows a few needs all prepared for use in product design. The needs, their translation, and their measurement are all placed to the left of the spreadsheet. The remainder of the spreadsheet will be discussed in the next section.

Step 4—Design: The Product or Service

Once the customers and their needs are fully understood, we are ready to design the organization. Most organizations have some process for designing and bringing new products to market. In this step of the Quality by Design process, we will focus on the role of quality in product development and how that role combines with the technical aspects of development and design appropriate for a particular industry. Within product development,

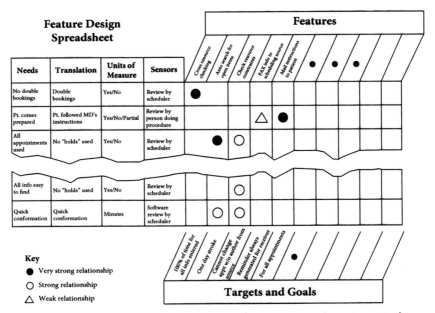

Figure 5.4 Product design spreadsheet. (From Juran Institute, Inc., 2013)

product design is a creative process based largely on technological or functional expertise.

The designers of products traditionally have been engineers, systems analysts, operating managers, and many other professionals. In the quality arena, designers can include any whose experience, position, and expertise can contribute to the design process. The outputs of product design are detailed designs, drawings, models, procedures, specifications, and so on.

There are two overall quality objectives for this step:

1. Determine which features and goals will provide the optimal benefit for the customer.
2. Identify what is needed so that the designs can be delivered without failures.

In the case of designing services, the scope of this activity is sometimes puzzling. For example, in delivering health care, where does the product of diagnosing and treating end and the processes of laboratory testing, chart reviews, and so on begin? One useful way to think about the distinction is that the product is the "face to the customer." It is what the customer sees and experiences. The patient sees and experiences the physician interaction, waiting time, clarity of information, and so on. The effectiveness and efficiency of moving blood samples to and around the laboratory have an effect on these features but are really features of the process that delivers the ultimate product to the customer.

Those who are designing physical products also can benefit from thinking about the scope of product design. Given that the customer's needs are the benefits that the customer wants from the product, the design of a piece of consumer electronics includes not only the contents of the box itself but also the instructions for installation and use and the help line for assistance. There are six major activities in this step:

1. Group together related customer needs.
2. Determine methods for identifying features.
3. Select high-level features and goals.
4. Develop detailed features and goals.
5. Optimize features and goals.
6. Set and publish final product design.

Group Together Related Customer Needs

Most Quality by Design projects will be confronted with a large number of customer needs. Based on the data developed in the preceding steps, the team can prioritize and group together those needs that relate to similar functionality. This activity does not require much time, but it can save a lot of time later. Prioritization ensures that the scarce resources of product development are spent most effectively on those items that are most important to the customer. Grouping related needs together allows the design team to "divide and conquer," with subteams working on different parts of the design. Such subsystem or component approaches to design, of course, have been common for years. What may be different here is that the initial focus is on the components of the customers' needs, not the components of the product. The component design for the product will come during the later activities in this step.

Determine Methods for Identifying Features

There are many complementary approaches for identifying the best product design for meeting customers' needs. Most design projects do not use all of them. Before starting to design, however, a team should develop a systematic plan for the methods it will use in its own design. Here are some of the options.

Benchmarking   This approach identifies the best in class and the methods behind it that make it best.

Basic Research   One aspect of research might be a new innovation for the product that does not currently exist in the market or with competitors. Another aspect of basic research looks at exploring the feasibility of the product and features. While both these aspects are important, be careful that fascination with the technological abilities of the product does not overwhelm the primary concern of its benefits to the customer.

Market Experiments   Introducing and testing ideas for features in the market allows one to analyze and evaluate concepts. The focus group is one technique that can be used to measure customer reactions and determine whether the features actually will meet customer needs. Some organizations also try out their ideas, on an informal basis, with customers at trade shows and association meetings. Still others conduct limited test marketing with a prototype product.

Creativity   Developing features allows one to dream about a whole range of possibilities without being hampered by any restrictions or preconceived notions. Design for quality is a proven, structured, data-based approach to meeting customers' needs. But this does not mean it is rigid and uncreative. At this point in the process, the participants in design must be encouraged and given the tools they need to be creative so as to develop alternatives for design. After they have selected a number of promising alternatives, they will use hard analysis and data to design the final product.

Design teams can take advantage of how individuals view the world: from their own perspective. Every employee potentially sees other ways of doing things. The team can encourage people to suggest new ideas and take risks. Team members should avoid getting "stuck" or taking too much time to debate one particular idea or issue. They can put it aside and come back to it later with a fresh viewpoint. They can apply new methods of thinking about customers' needs or problems, such as the following:

- *Change in key words or phrases.* For example, call a "need" or "problem" an "opportunity." Instead of saying, "Deliver on time," say, "Deliver exactly when needed."
- *Random association.* For example, take a common word such as *apple* or *circus* and describe your business, product, or problem as the word. For example, "Our product is like a circus because . . ."
- *Central idea.* Shift your thinking away from one central idea to a different one. For example, shift the focus from the product to the customer by saying, "What harm might a child suffer, and how can we avoid it?" rather than "How can we make the toy safer?"
- *Putting yourself in the other person's shoes.* Examine the question from the viewpoint of the other person, your competitor, your customer—and build their case before you build your own.
- *Dreaming.* Imagine that you had a magic wand that you could wave to remove all obstacles to achieving your objectives. What would it look like? What would you do first? How would it change your approach?
- *The spaghetti principle.* When you have difficulty considering a new concept or how to respond to a particular need, allow your team to be comfortable enough to throw out a new idea, as if you were throwing spaghetti against the wall, and see what sticks. Often even "wild" ideas can lead to workable solutions.

The initial design decisions are kept as simple as possible at this point. For example, the idea of placing the control panel for the radio on the steering wheel would be considered a high-level product feature. Its exact location, choice of controls, and how they function can be analyzed later in greater detail. It may become the subject of more detailed features as the design project progresses.

Standards, Regulations, and Policies   This is also the time to be certain that all relevant standards, regulations, and policies have been identified and addressed. While some of these requirements are guidelines for how a particular product or product feature can perform, others mandate how they must perform. These may come from inside the organization, and others may come from specific federal, state, or local governments; regulatory agencies; or industry associations. All features and product feature goals must be analyzed against these requirements prior to the final selection of features to be included in the design.

It is important to note that if there is a conflict when evaluating features against any standards, policies, or regulations, it is not always a reason to give up. Sometimes one can work to gain acceptance for a change when it will do a better job of meeting customer needs. This is especially true when it comes to internal policies. However, an advocate for change must be prepared to back up the arguments with appropriate data.

Criteria for Design   As part of the preparation for high-level design, the design team must agree on the explicit criteria to be used in evaluating alternative designs and design features. All designs must fulfill the following general criteria:

- Meet the customers' needs
- Meet the suppliers' and producers' needs
- Meet (or beat) the competition
- Optimize the combined costs of the customers and suppliers

In addition to the preceding four general criteria, the team members should agree explicitly on the criteria that they will use to make a selection. (If the choices are relatively complex, the team should consider using the formal discipline of a selection matrix.) One source for these criteria will be the team's goal statement and goals. Some other types of criteria that the team may develop could include:

- The impact of the feature on the needs
- The relative importance of the needs being served
- The relative importance of the customers whose needs are affected
- The feasibility and risks of the proposed feature
- The impact on product cost
- The relationship to competitive features uncovered in benchmarking
- The requirements of standards, policies, regulations, mandates, and so on

As part of the decision on how to proceed with design, teams also must consider a number of other important issues regarding what type of product feature will be the best response to customers' needs. When selecting features, they need to consider whether to:

- Develop an entirely new functionality
- Replace selected old features with new ones

- Improve or modify existing features
- Eliminate the unnecessary

Select High-Level Features and Goals

This phase of Quality by Design will stimulate the team to consider a whole array of potential features and how each would respond to the needs of the customer. This activity should be performed without being constrained by prior assumptions or notions as to what worked or did not work in the past. A response that previously failed to address a customer need or solve a customer problem might be ready to be considered again because of changes in technology or the market.

The team begins by executing its plan for identifying the possible features. It should then apply its explicit selection criteria to identify the most promising features.

The product design spreadsheet in Figure 5.4 is a good guide for this effort. Use the right side of the spreadsheet to determine and document the following:

- Which features contribute to meeting which customer needs
- That each priority customer need is addressed by at least one product feature
- That the total impact of the features associated with a customer need is likely to be sufficient for meeting that need
- That every product feature contributes to meeting at least one significant customer need
- That every product feature is necessary for meeting at least one significant customer need (i.e., removing that feature would leave a significant need unmet)

Team Sets Goals for Each Feature   In quality terms, a goal is an aimed-at quality target (such as aimed-at values and specification limits). As discussed earlier, this differs from quality standards in that the standard is a mandated model to be followed that typically comes from an external source. While these standards serve as "requirements" that usually dictate uniformity or how the product is to function, product feature goals are often voluntary or negotiated. Therefore, the Quality by Design process must provide the means for meeting both quality standards and quality goals.

Criteria for Setting Product Feature Goals As with all goals, product feature goals must meet certain criteria. While the criteria for establishing product feature goals differ slightly from the criteria for project goals verified in step 1, there are many similarities. Product feature goals should encompass all the important cases and be:

- Measurable
- Optimal
- Legitimate
- Understandable
- Applicable
- Attainable

Develop Detailed Features and Goals

For large and highly complex products, it will usually be necessary to divide the product into a number of components and even subcomponents for detailed design. Each component will typically have its own design team that will complete the detailed design described below. To ensure that the overall design remains integrated, consistent, and effective in meeting customer needs, these large, decentralized projects require:

- A steering or core team that provides overall direction and integration
- Explicit charters with quantified goals for each component
- Regular integrated design reviews for all components
- Explicit integration of designs before completion of the product design phase

Once the initial detailed features and goals have been developed, then the technical designers will prepare a preliminary design, with detailed specifications. This is a necessary step before a team can optimize models of features using a number of Quality by Design tools and ultimately set and publish the final features and goals.

It is not uncommon for Quality by Design teams to select features at so high a level that the features are not specific enough to respond to precise customer needs. Just as in the identification of customers' primary needs, high-level features need to be broken down further into terms that are clearly defined and can be measured.

Optimize Features and Goals

Once the preliminary design is complete, it must be optimized. That is, the design must be adjusted so that it meets the needs of both customer and supplier while minimizing their combined costs and meeting or beating the competition.

Finding the optimum can be a complicated matter unless it is approached in an organized fashion and follows quality disciplines. For example, there are many designs in which numerous variables converge to produce a final result. Some of these designs are of a business nature, such as design of an information system involving optimal use of facilities, personnel, energy, capital, and so on. Other such designs are technological, involving optimization of the performance of hardware. Either way, finding the optimum is made easier through the use of certain quality disciplines.

Finding the optimum involves balancing the needs, whether they are multiorganizational needs or within-organization needs. Ideally, the search for the optimum should be done through the participation of suppliers and customers alike. There are several techniques that help achieve this optimum.

Design Review Under this concept, those who will be affected by the product are given the opportunity to review the design during various formative stages. This allows them to use their experience and expertise to make such contributions as:

- Early warning of upcoming problems
- Data to aid in finding the optimum
- Challenge to theories and assumptions

Design reviews can take place at different stages of development of the new product. They can be used to review conclusions about customer needs and hence the product specifications (characteristics of product output). Design reviews also can take place at the time of selecting the optimal product design. Typical characteristics of design reviews include the following:

- Participation is mandatory
- Reviews are conducted by specialists, external to the design team
- Ultimate decisions for changes remain with the design team
- Reviews are formal, scheduled, and prepared for with agendas

- Reviews will be based on clear criteria and predetermined parameters
- Reviews can be held at various stages of the project

Ground rules for good design reviews include:

- Adequate advance design review of agenda and documents
- Clearly defined meeting structure and roles
- Recognition of interdepartmental conflicts in advance
- Emphasis on constructive—not critical—inputs
- Avoidance of competitive design during review
- Realistic timing and schedules for the reviews
- Sufficient skills and resources provided for the review
- Discussion focus on untried/unproved design ideas
- Participation directed by management

Multifunctional Design Teams   Design teams should include all those who have a vested interest in the outcome of the design of the product along with individuals skilled in product design. Under this concept, the team members, rather than just the product designers, bear responsibility for the final design.

Set and Publish Final Product Design

After the design has been optimized and tested, it is time to select the features and goals to be included in the final design. This is also the stage where the results of product development are officially transmitted to other functions through various forms of documentation. These include the specifications for the features and product feature goals as well as the spreadsheets and other supporting documents. All this is supplemented by instructions, both oral and written. To complete this activity, the team must first determine the process for authorizing and publishing features and product feature goals. Along with the features and goals, the team should include any procedures, specifications, flow diagrams, and other spreadsheets that relate to the final product design. The team should pass along results of experiments, field testing, prototypes, and so on that are appropriate. If an organization has an existing process for authorizing product goals, it should be reexamined in light of recent experience. Ask these questions: Does the authorization process guarantee input from key customers—both internal and external? Does it provide for optimization of the design? If an organization has no existing goal authorization process, now is a good time to initiate one.

Step 5—Develop: The Process

Once the product is designed and developed, it is necessary to determine the means by which the product will be created and delivered on a continuing basis. These means are, collectively, the *process*. *Process development* is the set of activities for defining the specific means to be used by operating personnel for meeting product quality goals. Some related concepts include:

- *Subprocesses*: Large processes may be decomposed into these smaller units for both the development and operation of the process.
- *Activities*: These are steps in a process or subprocess.
- *Tasks*: These comprise detailed, step-by-step description for execution of an activity.

For a process to be effective, it must be goal-oriented, with specific measurable outcomes; systematic, with the sequence of activities and tasks fully and clearly defined and all inputs and outputs fully specified; capable, i.e., able to meet product quality goals under operating conditions; and legitimate, with clear authority and accountability for its operation.

The 11 major activities involved in developing a process are as follows:

1. Review product goals.
2. Identify operating conditions.
3. Collect known information on alternate processes.
4. Select general process design.
5. Identify process features and goals.
6. Identify detailed process features and goals.
7. Design for critical factors and human error.
8. Optimize process features and goals.
9. Establish process capability.
10. Set and publish final process features and goals.
11. Set and publish final process design.

User's Understanding of the Process

By *users*, we mean either those who contribute to the processes in order to meet product goals or those who employ the process to meet their own needs. Users consist, in part, of internal customers (organization units or persons) responsible for running the processes to meet the quality goals. Operators or other workers are users. Process planners need to know how these people will understand the work to be done. The process must be designed either to accommodate this level of understanding or to improve the level of understanding.

How the Process Will Be Used

Designers always know the intended use of the process they develop. However, they may not necessarily know how the process is actually used (and misused) by the end user. Designers can draw on their own experiences but usually must supplement these with direct observation and interviews with those affected.

Identify Process Features and Controls

A *process feature* is any property, attribute, and so on that is needed to create the goods or deliver the service and achieve the product feature goals that will satisfy a customer need. A *process goal* is the numeric target for one of the features.

Whereas features answer the question "What characteristics of the product do we need to meet customers' needs?" process features answer the question "What mechanisms do we need to create or deliver those characteristics (and meet quality goals) over and over without failures?" Collectively, process features define a process. The flow diagram is the source of many of, but not all, these features and goals.

As the process design progresses from the macro level down into details, a long list of specific process features emerges. Each of these is aimed directly at producing one or more features. For example:

- Creating an invoice requires a process feature that can perform arithmetic calculations so that accurate information can be added
- Manufacturing a gear wheel requires a process feature that can bore precise holes into the center of the gear blank
- Selling a credit card through telemarketing requires a process feature that accurately collects customer information

Most process features fall into one of the following categories:

- Procedures—a series of steps followed in a regular, definite order

- ▲ Methods—an orderly arrangement of a series of tasks, activities, or procedures
- ▲ Equipment and supplies—"physical" devices and other hard goods that will be needed to perform the process
- ▲ Materials—tangible elements, data, facts, figures, or information (these, along with equipment and supplies, also may make up inputs required as well as what is to be done to them)
- ▲ People—numbers of individuals, skills they will require, goals, and tasks they will perform
- ▲ Training—skills and knowledge required to complete the process
- ▲ Other resources—additional resources that may be needed
- ▲ Support processes—secretarial support, occasionally other support, such as outsources of printing services, copying services, temporary help, and so on

Just as in the case of product design, process design is easier to manage and optimize if the process features and goals are organized into a spreadsheet indicating how the process delivers the features and goals. Figure 5.5 illustrates such a spreadsheet.

The spreadsheet serves not only as a convenient summary of the key attributes of the process, it also facilitates answering two key questions that are necessary for effective and efficient process design. First, will every

Product Feature	Product Feature Goal	Process Features			
		Spray delivery capacity	Crew size	Certified materials	Scheduling forecast on PC to determine to/from and work needed
Time to perform job	Less than one hour 100 pecent of the time	O	●		●
Guaranteed appointment time	99 percent of jobs within 15 min. of appointment				●
All materials enviromentally safe	All naturally occurring/ no synthetics			●	
Legend ● Very Strong O Strong △ Weak		10 gallons per minute	One person per 10,000 sq. ft. of yd.	100% approved by State Dept. of Agriculture	Forecast time always within 10 percent of actual
		Process Feature Goals			

Figure 5.5 Process design spreadsheet. (From Juran Institute, Inc., 1994)

product feature and goal be attained by the process? Second, is each process feature absolutely necessary for at least one product feature; i.e., are there any unnecessary or redundant process features? Also, verify that one of the other process features cannot be used to create the same effect on the product.

Often high-level process designs will identify features and goals that are required from organizationwide macro processes. Examples might include cycle times from the purchasing process, specific data from financial systems, and new skills training. Because the new process will depend on these macro processes for support, now is the time to verify that they are capable of meeting the goals. If they are not, the macro processes will need to be improved as part of the process design, or they will need to be replaced with an alternative delivery method.

Optimize Process Features and Goals

After the planners have designed for critical factors and made modifications to the plan for ways of reducing human error, the next activity is to optimize first the subprocesses and then the overall process design. In step 4, develop product, the concept of optimization was introduced. The same activities performed for optimizing features and product feature goals also apply to process planning. Optimization applies to the design of both the overall process and the individual subprocesses.

Establish Process Capability

Before a process begins operation, it must be demonstrated to be capable of meeting its quality goals. Any design project must measure the capability of its process with respect to the key quality goals. Failure to achieve process capability should be followed by systematic diagnosis of the root causes of the failure and improvement of the process to eliminate those root causes before the process becomes operational.

Set and Publish Final Process Features and Goals

After the design team has established the flow of the process, identified initial process features and goals, designed for critical processes and human error, optimized process features and goals, and established process capabilities, it is ready to define all the detailed process features and goals to be included in the final design. This is also the stage where the results of process development are officially transmitted to other functions through various forms of documentation. These include the specifications for the features and product feature goals as well as the spreadsheets and other supporting documents. All this is supplemented by instructions, both oral and written.

Step 6—Deliver: The Transfer to Operations

In this step, planners develop controls for the processes, arrange to transfer the entire product plan to operational forces, and validate the implementation of the transfer. There are seven major activities in this step:

1. Identify controls needed.
2. Design feedback loop.
3. Optimize self-control and self-inspection.
4. Establish an audit.
5. Demonstrate process capability and controllability.
6. Plan for transfer to operations.
7. Implement plan and validate transfer.

Once design is complete, these plans are placed in the hands of the operating departments. It then becomes the responsibility of the operational personnel to manufacture the goods or deliver the service and to ensure that quality goals are met precisely and accurately. They do this through a planned system of quality control. Control is largely directed toward continuously meeting goals and preventing adverse changes from affecting the quality of the product. Another way of saying this is that no matter what takes place during production (change or loss of personnel, equipment or electrical failure, changes in suppliers, etc.), workers will be able to adjust or adapt the process to these changes or variations to ensure that quality goals can be achieved.

Identify Controls Needed

Process control consists of three basic activities:

- Evaluate the actual performance of the process
- Compare actual performance with the goals
- Take action on the difference

Demonstrate Process Capability and Controllability

While process capability must be addressed during the design of the process, it is during implementation that initial findings of process capability and controllability must be verified.

Plan for Transfer to Operations

In many organizations, receipt of the process by operations is structured and formalized. An information package is prepared consisting of certain standardized essentials: goals to be met, facilities to be used, procedures to be followed, instructions, cautions, and so on. There are also supplements unique to the project. In addition, provision is made for briefing and training the operating forces in such areas as maintenance, dealing with crisis, and so on. The package is accompanied by a formal document of transfer of responsibility. In some organizations, this transfer takes place in a near-ceremonial atmosphere.

Implement Plan and Validate Transfer

The final activity of the Quality by Design process is to implement the plan and validate that the transfer has occurred. A great deal of time and effort has gone into creating the product plan, and validating that it all works is well worth the effort.

References

Designs for World Class Quality. (1995). Juran Institute, Wilton, Conn.
Juran, J. M. (1992). *Quality by Design.* Free Press, New York.
Juran, J. M. (1999). *Quality Control Handbook,* 5th ed. McGraw-Hill, New York., p. 3.12.
Parasuraman, A., V. A. Zeithami, and L. L. Berry. (1985). "A Conceptual Model for Service Quality and Its Implications for Further Research." *Journal of Marketing,* Fall, pp. 41–50.
Quality by Design. (2013). Juran Institute, Southbury, Conn.

CHAPTER 6

Creating Breakthroughs in Performance

The purpose of this chapter is to explain the means to achieve breakthroughs and its relation to attaining superior results. This chapter deals with the universal and fundamental concepts that define the methods to create *breakthroughs in current performance*. The Six Sigma Model for Performance Improvement, popularized by Motorola and GE, is the most widely used method for attaining breakthrough.

The Universal Sequence for Breakthrough

Improvement happens every day, in every organization—even among the poor performers. That is how businesses survive—in the short term. Improvement is an activity in which every organization carries out tasks to make incremental improvements, day after day. Improvement is different from breakthrough improvement. Breakthrough requires special methods and support to attain significant changes and results. It also differs from planning and control. Breakthrough requires taking a "step back" to discover what may be preventing the current level of performance from meeting the needs of its customers.

As used here, *breakthrough* means "the organized creation of beneficial change and the attainment of unprecedented levels of performance." Synonyms are *quality improvement* or *Six Sigma improvement*. Unprecedented change may require attaining a Six Sigma level (3.4 ppm) or tenfold levels of improvement over current levels of process performance. Breakthrough results in significant cost reduction, customer satisfaction enhancement, and superior results that will satisfy stakeholders.

The concept of a universal sequence evolved from Dr. Juran's experience first in Western Electric Organization (1924–1941) and later during my years as an independent consultant, starting in 1945. Following a few preliminary published papers, a universal sequence was published in book form (Juran, 1964). This sequence then continued to evolve based on experience gained from applications by operating managers.

Breakthrough means change—a dynamic, decisive movement to new, higher levels of performance. In a truly static society, breakthrough is taboo, forbidden. There have been many such societies, and some have endured for centuries. During those centuries, their members either suffered or enjoyed complete predictability. They knew precisely what their station in life was—the same as that lived out by their forebears—but this predictability was, in due course, paid for by a later generation. The price paid was the extinction of the static society through conquest or another takeover by some form of society that was on the move. The threat of extinction may well have been known to the leaders of some of these static societies. Some gambled that the threat would not become a reality until they were gone. It was well stated in Madame de Pompadour's famous letter to Louis XV of France: "After us, the deluge."

Breakthrough is applicable to any industry, problem, or process. To better understand why so many organizations create extensive quality improvement programs such as Lean Six Sigma, we must contrast planning versus improvement. In Chapter 5, we discussed the quality planning process to design features.

Breakthrough to reduce excess failures and deficiencies may consist of such actions as these:

- Increase the yield of production processes.
- Reduce error rates of administrative reports.
- Reduce field failures.
- Reduce claim denials.
- Reduce the time it takes to perform critical patient clinical procedures.

The experience of recent decades has led to an emerging consensus that managing for quality (planning, control, and improvement) is one of the most cost-effective means to deal with the threats and opportunities, and to provide a means of action that needs to be taken. As it relates to breakthrough, the high points of this consensus include the following:

- Global competition has intensified and has become a permanent unpleasant fact. A needed response is to create a high rate of breakthrough, year after year.
- Customers are increasingly demanding improved products from their suppliers. These demands are then transmitted through the entire supplier chain. The demands may go beyond product breakthrough and extend to improving the system of managing for quality.
- The chronic wastes can be huge in organizations that do not have a strategic program aimed at reducing them. In many organizations during the early 1980s, about one-third of all work consisted of redoing what was done previously, due to deficiencies. By the end of the 1990s, this number improved to only 20 to 25 percent (estimated by the authors). The emerging consensus is that such waste should not continue, since it reduces competitiveness and profitability.
- Breakthroughs must be directed at all areas that influence an organization's performance: all business, transactional, and manufacturing processes.
- Breakthroughs should not be left solely to voluntary initiatives; they should be built into the strategic plan and DNA of a system. They must be mandated.
- Attainment of market leadership requires that the upper managers personally take charge of managing for quality. In organizations that did attain market leadership, the upper managers personally guided the initiative. The authors are not aware of any exceptions.

Unstructured Reduction of Chronic Waste

In most organizations, the urge to reduce costs has been much lower than the urge to increase sales. As a result:

- The business plan has not included goals for the reduction of chronic waste.
- Responsibility for such breakthroughs has been vague. It has been left to volunteers to initiate action.
- The needed resources have not been provided, since such breakthroughs have not been a part of the business plan.

The lack of priority by upper managers is traceable in large part to two factors that influence the thinking processes of many upper managers:

- Not only do many upper managers give top priority to increasing sales, but some of them even regard cost reduction as a form of lower-priority work that is not worthy of the time of upper managers. This is especially the case in high-tech industries.
- Upper managers have not been aware of the size of the chronic waste, or of the associated potential for high return on investment. The "instrument panel or scorecards" available to upper managers have stressed performance measures such as sales, profit, cash flow, and so on, but not the size of chronic waste and the associated opportunities. The managers have contributed to this unawareness by presenting their reports in the language of specialists rather than in the language of management—the language of money.

Breakthrough Models and Methods

Breakthrough addresses the question, How do I reduce or eliminate things that are wrong with my products, services, or processes and the associated customer dissatisfaction? Breakthrough models must address problems that create customer dissatisfaction, products and services of poor quality, and failures to meet the specific needs of specific customers, internal and external.

Based on Dr. Juran's research, attaining breakthroughs in current performance by reducing customer-related problems has one of the greatest returns on investment and usually comes down to correcting just a few types of things that go wrong, including these:

- Excessive number of defects
- Excessive numbers of delays or excessively long time cycles
- Excessive costs of the resulting rework, scrap, late deliveries, dealing with dissatisfied customers, replacement of returned goods, loss of customers and clients, loss of goodwill, etc.
- High costs and ultimately high prices, due to the waste

Effective breakthrough models require that:

- Leaders mandate it, project by project year after year

- Projects be assigned to teams that must discover root causes of the problems to sustain the gains
- Teams devise remedial changes to the "guilty" processes to remove or deal with the cause(s)
- Teams work with functions to install new controls to prevent the return of the causes
- Teams look for ways to replicate the remedies to increase the effect of the breakthrough
- All teams follow a systematic fact-based method like that used in Six Sigma, which requires making two journeys:
  - *The diagnostic journey* from symptoms (evidence a problem exists) to theories about what may cause the symptom(s); from theories to testing of the theories; from tests to establishing root cause(s). Once the causes are found, a second journey takes place.
  - *The remedial journey* from root causes to remedial changes in the process to remove or deal with the cause(s); from remedies to testing and proving the remedies under operating conditions; from workable remedies to dealing with resistance to change; from dealing with resistance to establishing new controls to hold the gains.
- Regardless of what your organization calls or brands its improvement model, breakthrough results only occur after the completion of both journeys.

The most recent success is Six Sigma or Six Sigma DMAIC. Six Sigma has become the most effective "brand" of improvement since Motorola Corporation first began using the quality improvement method Dr. Juran espoused in the late 1970s. Six Sigma methods and tools employ many of these universal principles. They have been combined with the rigor of statistical and technological tools to collect and analyze data.

GE's former chairman, Jack Welch, defined Six Sigma in this way: "Six Sigma is a quality program that, when all is said and done, improves your customers' experiences, lowers your costs and builds better leaders" (Welch, 2005).

Breakthrough Lessons Learned

My analysis of the actions taken by the successful organizations shows that most of them carried out many of or all the following tasks or strategies:

1. They enlarged the business plan at all levels to include annual goals for breakthrough and customer satisfaction.
2. They implemented a systematic process for making breakthroughs and set up special infrastructure or organizational machinery to carry out that process.
3. They adopted the big Q concept—they applied the breakthrough methods to all business processes, not just the manufacturing processes.
4. They trained all levels of personnel, including upper management, in the methods and tools to carry out their respective goals.
5. They enabled the workforce to participate in making breakthroughs in their daily work practices.
6. They established measures and scorecards to evaluate progress against the breakthrough goals.
7. The managers, including the upper managers, reviewed progress against the breakthrough goals.
8. They expanded the use of recognition and revised the reward system to recognize the changes in job responsibilities and use of the new methods and tools.
9. They renewed their programs every few years to include changes to their programs as performance improved.
10. They created a *rate of improvement* that exceeded the competition's.

The Rate of Breakthrough Is Most Important

The tenth lesson learned is an important one. Just having a system of breakthrough may not be enough. This lesson learned demonstrated that the annual rate of breakthrough determines which organizations will emerge as the market leaders. Figure 6.1 shows the effect of differing rates of breakthrough.

In this figure, the vertical scale represents product salability, so what goes up is good. The upper line shows the performance of company A, which at the outset was the industry leader. Company A kept getting better, year after year. In addition, company A was profitable. Company A seemed to face a bright future.

The lower line shows that company B, a competitor, was at the outset not the leader. However, company B is improving at a rate much faster than

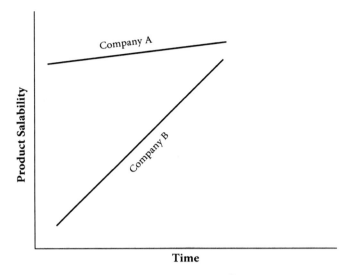

Figure 6.1 Two contrasting rates of improvement.
(From Juran Institute, Inc., 2013)

that of company A. Company A is now threatened with loss of its leadership when company B surpasses them. The lesson is clear:

> The most decisive factor in the competition for market leadership is the rate of breakthrough an organization maintains.
>
> JOSEPH M. JURAN

The sloping lines of Figure 6.1 help to explain why Japanese goods attained market leadership through quality in so many products. The major reason was that the Japanese organizations' rate of breakthrough was for decades revolutionary when compared with the evolutionary rate of the West. Eventually, they had to surpass the evolutionary rate of the Western organizations. The result was an economic disaster for many U.S. organizations in the early 1980s. Today, U.S. automobile manufacturers have made great strides in quality while Toyota has had recalls. Figure 6.2 shows my estimate of the rates of breakthrough in the automobile industry from 1950 to 1990.

There are also lessons to be learned from the numerous initiatives to improve competitiveness during the 1980s, some of which failed to pro-

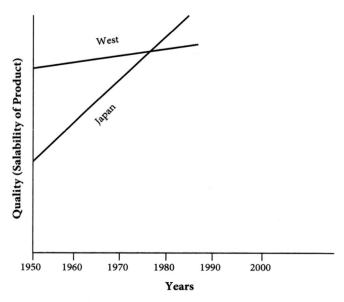

Figure 6.2 Estimate of rates of quality improvement in the automobile industry. (From Juran Institute, Inc., 2013)

duce bottom-line results. The quality circles, employee involvement teams, TQM, reengineering, and National Quality Awards all were methods used to respond to the Japanese quality revolution. Some were not sustainable and failed. Each of them may have helped the organization that used them at that time. An important lesson does stand out. The initiatives showed us that attaining a revolutionary rate of breakthrough is not simple at all. It takes a strategic focus to sustain market leadership. Only the National Quality Awards continue today in most parts of the world. Organizations that made statements such as "Quality is dead" or "TQM did not work" blamed the methodology for their failures. This was only partially true. In some cases, the wrong method was selected; in others, their own management did not deal with the numerous obstacles and cultural resistance that inhibited these methods from working in the first place. These obstacles and the means to manage them will be discussed throughout this chapter.

All Breakthrough Takes Place Project by Project

There is no such thing as breakthrough in a general way. All breakthrough takes place project by project and in no other way.

As used here, *breakthrough* means "the solving of a chronic problem by scheduling (launching a project) to find a solution." Since the word *breakthrough* has multiple meanings, the organization should create a glossary and educate all employees on what it means. The definition is reinforced by presenting a few examples that were carried out successfully in your organization.

The Backlog of Breakthrough Projects Is Never-Ending

The existence of a huge backlog of problems to solve is evident from the numbers of breakthroughs actually made by organizations that carried out successful initiatives during the 1980s and 1990s. Some reported making breakthroughs by the thousands, year after year. In very large organizations, the numbers are higher still, by orders of magnitude.

The backlog of breakthrough projects exists in part because of internal and external factors. Internally, the planning of new products and processes has long been deficient. In effect, the planning process has been a dual hatchery. It hatched out new plans. It also hatched out new chronic wastes, and these accumulated year after year. Each such chronic waste then became a potential breakthrough project.

A further reason for the huge backlog is the nature of human ingenuity—it seems to have no limit. Toyota Motor Corporation has reported that its 80,000 employees offered 4 million suggestions for breakthrough during a single year—an average of 50 suggestions per person per year (Sakai, 1994).

Externally, the constantly changing needs of customers and our society will always challenge the status quo. Targets today are not good enough for tomorrow. This creates a never-ending backlog of projects.

Breakthrough Does Not Come Free

Breakthrough and the resulting reduction of chronic waste do not come free—they require an expenditure of effort in several forms. It is necessary to create an infrastructure to mobilize the organization's resources toward annual breakthrough. This involves setting specific goals to be reached, choosing projects to be tackled, assigning responsibilities, following progress, and so on.

There is also a need to conduct extensive training in the nature of the breakthrough improvement methods and tools, how to serve on breakthrough teams, how to use the tools, and so on.

In addition to all this preparatory effort, each breakthrough improvement project requires added effort to conduct diagnoses to discover the causes of the chronic waste and provide remedies to eliminate the causes. This is the time it takes for all the people involved in the team to solve the problem.

The preceding adds up to a significant front-end outlay, but the results can be stunning. They *have* been stunning in the successful organizations, the role models. Detailed accounts of such results have been widely published, notably in the proceedings of the annual conferences held by the U.S. National Institute for Standards and Technology (NIST), which administers the Malcolm Baldrige National Award.

Reduction in Chronic Waste Is Not Capital-Intensive

Reduction in chronic waste seldom requires capital expenditures. Diagnosis to discover the causes usually consists of the time of the breakthrough project teams. Remedies to remove the causes usually involve fine-tuning the process. In most cases, a process that is already producing more than 80 percent good work can be raised to the high 90s without capital investment. Such avoidance of capital investment is a major reason why reduction of chronic waste has a high return on investment (ROI).

In contrast, projects to create breakthroughs in product design and development to increase sales can involve costly outlays to discover customer needs, design products and processes, build facilities, and so on. Such outlays are largely classified as capital expenditures and thereby lower the ROI estimates. There is also a time lag between investing in design and receiving revenue from the sale of the new designs.

The Return on Investment for Breakthrough Improvement Is High

This is evident from results publicly reported by national award winners in Japan (Deming Prize), the United States (Baldrige Award), Europe, and elsewhere. More and more organizations have been publishing reports describing their breakthroughs, including the gains made.

It has been noted that the actual return on investment from breakthrough projects has not been well researched. Dr. Juran's own research conducted by examining papers published by organizations found that the average breakthrough project yielded about $100,000 of cost reduction (Juran, 1985). The organizations were large—sales in the range of over $1 billion per year.

Dr. Juran had also estimated that for projects at the $100,000 level, the investment in diagnosis and remedy combined runs to about $15,000, or 15 percent. The resulting ROI is among the highest available to managers. It has caused some managers to quip, "The best business to be in is breakthrough." Today, breakthrough projects return many more dollars, but the cost of attaining breakthrough has not changed from the 15 percent investment level.

Dr. Juran was astounded by some of the recent organizations that have become world quality leaders using the project-by-project approach of Six Sigma. One of them is Samsung Electronics.

Samsung Electronics Co. (SEC) of Seoul, Korea, has perfected its fundamental improvement approach by using Six Sigma as a tool for innovation, efficiency, and quality. SEC was founded in 1969 and sold its first product, a television receiver, in 1971. Since that time, the company has used tools and techniques such as total quality control, total process management, product data management, enterprise resource management, supply-chain management, and customer relationship management. Six Sigma was added to upgrade these existing innovations and improve SEC's competitive position in world markets. The financial benefits made possible by Six Sigma, including cost savings and increased profits from sales and new product development, are expected to approach $1.5 billion.

SEC completed 3290 Six Sigma improvement projects in the first two years; 1512 of these were Black Belt–level projects. By the third year, 4720 projects are expected to be completed, 1640 of them by Black Belts.

SEC's Six Sigma projects have also contributed to an average of 50 percent reduction in defects. There is no thought of improvement in quality and productivity without Six Sigma. These impressive numbers have certainly played a major role in Samsung's recent growth. Some indications of this include the following:

- By 2001, SEC had earned a net income of $2.2 billion on total revenues of $24.4 billion. Market capitalization stood at $43.6 billion.

▲ According to SEC's 2001 annual report, SEC now is one of the top 10 electronic and electrical equipment manufacturing companies in the world, with the best operating profit ratios and superior fiscal soundness.
▲ The report also says the debt-to-equity ratio is lower than that of any top-ranking company, and the shareholders' equity-to-net-assets ratio surpasses the average.
▲ SEC says its technological strengths, Six Sigma quality initiatives, and product marketability helped increase its share of the memory chip market in 2001 to 29 percent, monitors to 21 percent, and microwave ovens to 25 percent of those sold worldwide.

Despite a downturn in the world economy and a reduction in exports to the United States, credit for SEC's current operating profit margin of 8.5 percent is due mostly to quality improvements and Six Sigma deployment.

SEC's quality and innovative strategy helped it reach the number-one position in the *BusinessWeek* 2002 information technology guide. The guide noted SEC's computer monitors, memory chips, telephone handsets, and other digital products, focusing on four Standard & Poor's criteria: shareholder return, return on equity, revenue growth, and total revenues.

The *BusinessWeek* ranking was also due to SEC employees' belief that quality is the single most important reason for the company's higher sales, lower costs, satisfied customers, and profitable growth. Only a few years ago, SEC's products were virtually unknown by Americans or were known as the cheaper, lower-quality substitute for Japanese brands. This perception is changing. The U.S. market now represents 37 percent of SEC's total sales.

The Major Gains Come from the Vital Few Projects

The bulk of the measurable gains comes from a minority of the breakthrough projects—the *vital few*. These are multifunctional in nature, so they need multifunctional teams to carry them out. In contrast, the majority of the projects are in the *useful many* category and are carried out by local departmental teams. Such projects typically produce results that are orders of magnitude smaller than those of the vital few.

While the useful many projects contribute only a minor part of the measurable gains, they provide an opportunity for the lower levels of the hierarchy, including the workforce, to participate in breakthrough. In the minds of many managers, the resulting gain in work life is as important as the tangible gains in operating performance.

Breakthrough—Some Inhibitors

While the role-model organizations achieved stunning results through breakthrough, most organizations did not. Some of these failures were due to honest ignorance of how to mobilize for breakthrough, but there are also some inherent inhibitors to establishing breakthrough on a year-to-year basis. It is useful to understand the nature of some of the principal inhibitors before setting out.

Disillusioned by the Failures

The lack of results mentioned earlier has led some influential journals to conclude that breakthrough initiatives are inherently doomed to failure. Such conclusions ignore the stunning results achieved by the role-model organizations. (Their results prove that these are achievable.) In addition, the role models have explained how they got those results, thereby providing lessons learned for other organizations to follow. Nevertheless, the conclusions of the media have made some upper managers wary about going into breakthrough.

The Illusion of Delegation

Managers are busy people, yet they are constantly bombarded with new demands on their time. They try to keep their workload in balance through delegation. The principle that "a good manager is a good delegator" has wide application, but it has been overdone as applied to breakthrough. The lessons learned from the role-model organizations show that going into annual breakthrough adds minimally about 10 percent to the workload of the entire management team, including the upper managers.

Most upper managers have tried to avoid this added workload through sweeping delegation. Some established vague goals and then exhorted everyone to do better—"Do it right the first time." In the role-model organizations, it was different. In every such organization, the upper managers took charge of the initiative and personally carried out certain nondelegable roles.

Employee Apprehensions

Going into breakthrough involves profound changes in a organization's way of life—far more than is evident on the surface. It adds new roles to the job descriptions and more work to the job holders. It requires accepting the concept of teams for tackling projects—a concept that is alien to many organizations and that invades the jurisdictions of the functional departments. It requires training on how to do all this. Collectively, the megachange disturbs the peace and breeds many unwanted side effects.

To the employees, the most frightening effect of this profound set of changes is the threat to jobs and/or status. Reduction of chronic waste reduces the need for redoing prior work and hence the jobs of people engaged in such rework. Elimination of such jobs then becomes a threat to the status and/or jobs of the associated supervision. It should come as no surprise if the efforts to reduce waste are resisted by the workforce, the union, the supervision, and others.

Nevertheless, breakthrough is essential to remaining competitive. Failure to go forward puts all jobs at risk. Therefore, the organization should go into breakthrough while realizing that employee apprehension is a logical reaction of worried people to worrisome proposals. A communication link must be opened to explain the why, understand the worries, and search for optimal solutions. In the absence of forthright communication, the informal channels take over, breeding suspicions and rumors.

Additional apprehension has its origin in cultural patterns. (The preceding apprehensions do not apply to breakthrough of product features to increase sales. These are welcomed as having the potential to provide new opportunities and greater job security.)

Securing Upper Management Approval and Participation

The lessons learned during the 1980s and 1990s included a major finding: Personal participation by upper managers is indispensable to getting a high rate of annual breakthrough. This finding suggests that advocates for initiatives should take positive steps to convince the upper managers of the following:

▲ The merits of planning for annual breakthrough

▲ The need for active upper management to provide resources
▲ The precise nature of the needed upper management participation

Proof of the Need

Upper managers respond best when they are shown a major threat or opportunity. An example of a major threat is seen in the case of organization G, a maker of household appliances. Organization G and its competitors, R and T, were all suppliers to a major customer involving four models of appliances. (See Table 6.1.) This table shows that in 2000, organization G was a supplier for two of the four models. Organization G was competitive in price, on-time delivery, and product features, but it was definitely inferior in the customer's perception of the chief problem, field failures. By 2002, lack of response had cost organization G the business on model number 1. By 2003, organization G also had lost the business on model number 3.

Table 6.1 Suppliers to a Major Customer

Model Number	2000	2001	2002	2003
1	G	G	R	R
2	R	R	R	R
3	G	G	G	R
4	T	R	R	R

Awareness also can be created by showing upper managers other opportunities, such as cost reduction through cutting chronic waste.

The Size of the Chronic Waste

A widespread major opportunity for upper managers is to reduce the cost of poor quality or the costs associated with poorly performing processes. In most cases, this cost is greater than the organization's annual profit, often much greater. Quantifying this cost can go far toward proving the need for a radical change in the approach to breakthrough. An example is shown in Table 6.2. This table shows the estimated cost of poor quality (COPQ) for an organization in a process industry using the traditional accounting classifications. The table brings out several matters of importance to upper managers.

Table 6.2 Analysis of Cost of Poor Quality

Category	Amount, $	Percentage of Total
Internal failures	7,279,000	79.4
External failures	283,000	3.1
Appraisal	1,430,000	15.6
Prevention	170,000	1.9
	9,162,000	100.0

- *The order of magnitude.* The total of the costs is estimated at $9.2 million per year. For this organization, this sum represented a major opportunity. (When such costs have never before been brought together, the total is usually much larger than anyone would have expected.)
- *The areas of concentration.* The table is dominated by the costs of internal failures—they are 79.4 percent of the total. Clearly, any major cost reduction must come from the internal failures.

COPQ versus Cost Reduction

Company X wanted to reduce operating costs by 10 percent. It began with a mission to have each executive identify where costs could be cut in business units. The executives created a list of 60 items, including things such as eliminating quality audits, changing suppliers, adding new computer systems, reducing staff in customer services, and cutting back R&D.

The executives removed functions that provide quality and services to meet customer needs. They bought inferior parts and replaced computer systems at great expense. They disrupted their organization, particularly where the customers were most affected, and reduced the potential for new services in the future.

After accomplishing this, most of the executives were rewarded for their achievements. The result? Their cost reduction goal was met, but they had dissatisfied employees, upset customers, and an organization that still had a significant amount of expense caused by poor performance.

The financial benefit to the bottom line of an organization's balance sheet from improving the cost of quality is not always fully appreciated or understood. This misunderstanding stems from the old misconception that improving quality is expensive.

However, this misconception is partially true. For example, if your organization provides a service to clients for a given price and a competitor provides the same basic service with enhanced features for the same price, it will cost your organization more to add those features that the competitor already provides.

If your organization does not add those features, it may lose revenue because customers will go to a competitor. If you counteract by reducing the price, you may still lose revenue. In other words, the quality of your competitor's service is better.

For your organization to remain competitive, it will have to invest in developing new features. This positively affects revenue. To improve quality, features have to be designed in—or in today's terminology, a new design must be provided at high Sigma levels.

Because of this historical misconception, organizations do not always support the notion that improving quality will affect costs and not add to them. They overlook the enormous costs associated with poor performance of products, services, and processes—costs associated with not meeting customer requirements, not providing products or services on time, or reworking them to meet the customer needs. These are the costs of poor quality (COPQ) or the cost of poorly performing processes (COP^3).

If quantified, these costs will get immediate attention at all management levels. Why? When added together, costs of poor quality make up as much as 15 to 30 percent of all costs. Quality in this complete sense, unlike the quality that affects only income, affects costs. If we improve the performance of products, services, and processes by reducing deficiencies, we will reduce these costs. To improve the quality of deficiencies that exist throughout an organization, we must apply breakthrough improvements.

A Six Sigma program focused on reducing the costs of poor quality due to low Sigma levels of performance and on designing in new features (increasing the Sigma levels) will enable management to reap increased customer satisfaction and bottom-line results. Too many organizations reduce costs by eliminating essential product or service features that provide satisfaction to customers, while ignoring poor performance that costs the bottom line and shareholders millions of dollars.

A Better Approach

Company Y approached its situation differently than did company X, as described at the beginning of this section. The executives identified all costs that would disappear if everything worked better at higher Sigma levels. Their list included costs associated with credits or allowances given to customers because of late delivery, inaccuracy or errors in billings, scrap and rework, and accounts payable mistakes caused by discount errors and other mistakes.

When this company documented its costs of poor quality, the management team was astounded by the millions of dollars lost due to poor quality of performance within the organization.

This total cost of poor quality then became the target. The result? Elimination of waste and a return to the bottom line from planned cost reductions and more satisfied customers. Why? Because the company eliminated the reasons these costs existed in the first place. There were process and product deficiencies that caused customer dissatisfaction. Once these deficiencies were removed, the quality was higher and the costs were lower.

While it is becoming essential to respond to customer demands for improved quality in everything an organization does, organizations should not overlook the financial impact of poor performance. In fact, these costs should be the driver of the project selection process for Six Sigma.

In other words, the cost of poor quality provides proof of why changes must be made. The need to improve an organization's financial condition correlates directly with the process of making and measuring quality improvements. Regardless of the objective you start with, enhancing features as well as reducing costs of poor quality will affect the continuing financial success of an operation.

While there is a limit to the amount quality can be improved when cost effectiveness and savings are measured against the costs of achieving them, it's not likely this will occur until you approach Five or Six Sigma levels. A business must pursue the next level of quality based on what is of critical importance to its customers. If customers demand something, chances are it must be done to keep the business. If they do not, there's time to plan.

Driving Bottom-Line Performance

If you accept the reality that customers and the marketplace define quality, then your organization must have the right product or service features and lower your deficiencies to create loyal customers.

With a competitive price and market share strongly supported by fast cycle time, low warranty costs, and low scrap and rework costs, revenue will be higher and total cost lower. The substantial bonus that falls to the profit column comes, in effect, from a combination of enhancing features and reducing the costs of poor quality.

Before getting into specific ways to identify, measure, and account for the impact of costs of poor quality on financial results, you must look at what to do first if you are trying to understand how the costs of quality can drive a financial target.

For example, if your organization sets a cost reduction target to save $50 million, there is a simple methodology to determine how many improvement projects it will take to reach that goal. The organization can then manage the improvement initiative more effectively if it puts some thought behind how much activity it can afford. The answer will help determine how many experts or Black Belts are needed to manage the improvements and how much training will be required.

The methodology includes the following six steps:

1. Identify your cost reduction goal of $50 million over the next two years—$25 million per year.
2. Using an average return of $250,000 for each improvement, calculate the number of projects needed to meet the goal for each year. For this example, we would need an incredible 200 projects—100 per year.
3. Calculate how many projects per year can be completed and how many experts will be required to lead the team. If each project can be completed in four months, that means one Black Belt on two projects per four months. Hence, one Black Belt can complete six projects in one year. We will then need about 17 Black Belts.
4. Estimate how many employees will be involved on a part-time basis to work with the Black Belts to meet their targets. Assume four per Black Belt per four months. We would need about 200 employees involved at some level each year, possibly as little as 10 percent of their time.
5. Identify the specific costs related to poor performance, and select projects from this list that are already causing your organization to incur at least $250,000 per deficiency. If you haven't created this list, use a small team to identify the costs and create a Pareto analysis prior to launching any projects.

6. Use this method and debate each variable among the executive team to ensure the right amount of improvement can be supported. All organizations make improvements, but world-class organizations improve at a faster rate than their competition.

The Results

Of note is the fact that every organization that has adopted Six Sigma and integrated the discipline throughout its operations has produced impressive savings that were reflected on the bottom line. More customers were satisfied and became loyal, and revenues, earnings, and operating margins improved significantly.

For example, Honeywell's cost savings have exceeded $2 billion since implementing Six Sigma in 1994. At General Electric, the Six Sigma initiative began in 1996 and produced more than $2 billion in benefits in 1999. Black & Decker's Six Sigma productivity savings rose to about $75 million in 2000, more than double the prior year's level, bringing the total saved since 1997 to over $110 million.

A more revealing insight into the cost of poor quality as a function of Six Sigma performance levels is the following:

- When ± 3 Sigma of the process that produces a part is within specification, there will be 66,807 defects per million parts produced. If each defect costs $1000 to correct, then the total COPQ would be $66,807,000.
- When an organization improves the process to within ± 4 Sigma, there will be only 6210 defects per million at a COPQ of $6,210,000.
- At ± 5 Sigma, the cost of defects declines to $233,000 per million, a savings of $66,574,000 more than the savings at a process capability of ± 3 Sigma.
- At the near perfection level of ± 6 Sigma, defects are almost eliminated at $3400 per million parts produced.

The Potential Return on Investment

A major responsibility of upper managers is to make the best use of the organization's assets. A key measure of judging what is best is the return on investment (ROI). In general terms, ROI is the ratio of (1) the estimated gain to (2) the estimated resources needed. Computing ROI for projects to reduce chronic waste requires assembling estimates such as:

- The costs of chronic waste associated with the projects
- The potential cost reductions if the projects are successful
- The costs of the needed diagnosis and remedy

Many proposals to go into breakthrough have failed to gain management support because no one has quantified the ROI. Such a goal is a handicap to the upper managers—they are unable to compare (1) the potential ROI from breakthrough with (2) the potential ROI from other opportunities for investment.

Managers and others who prepare such proposals are well advised to prepare the information on ROI in collaboration with those who have expertise in the intricacies of ROI. Computation of ROI can be complicated because two kinds of money are involved—capital and expenses. Each is money, but in some countries (including the United States) they are taxed differently. Capital expenditures are made from after-tax money, whereas expenses are paid out of pretax money.

This difference in taxation is reflected in the rules of accounting. Expenses are written off promptly, thereby reducing the stated earnings and hence the income taxes on earnings. Capital expenditures are written off gradually—usually over a period of years. This increases the stated earnings and hence the income taxes on those earnings. This means it is advantageous for proposals to go into breakthrough because breakthrough is seldom capital-intensive. (Some upper managers tend to use the word *investment* as applying only to capital investment.)

Upper managers receive numerous proposals for allocating the organization's resources: invade foreign markets, develop new products, buy new equipment to increase productivity, make acquisitions, enter into joint ventures, and so on. These proposals compete with one another for priority, and a major test is ROI. It helps if the proposal to go into breakthrough includes estimates of ROI.

An explanation of proposals is sometimes helped by converting the supporting data into units of measure that are already familiar to upper managers. For example:

- Last year's cost of poor quality was five times last year's profit of $1.5 million.
- Cutting the cost of poor quality in half would increase earnings by 13 cents per share of stock.
- Thirteen percent of last year's sales orders were canceled due to poor quality.
- Thirty-two percent of engineering time was spent in finding and correcting design weaknesses.
- Twenty-five percent of manufacturing capacity is devoted to correcting problems.
- Seventy percent of the inventory carried is traceable to poor quality.
- Twenty-five percent of all manufacturing hours were spent in finding and correcting defects.
- Last year's cost of poor quality was the equivalent of our operation making 100 percent defective work during the entire year.

Experience in making presentations to upper management has provided some useful dos and don'ts:

- *Do* summarize the total of the estimated costs of poor quality. The total will be big enough to command upper management's attention.
- *Do* show where these costs are concentrated. A common grouping is in the form of Table 6.2. Typically (as in that case), most of the costs are associated with failures, internal and external. Table 6.2 also shows the fallacy of trying to start by reducing inspection and test. The failure costs should be reduced first. After the defect levels come down, inspection costs can be reduced as well.
- *Do* describe the principal projects that are at the heart of the proposal.
- *Do* estimate the potential gains, as well as the return on investment. If the organization has never before undertaken an organized approach to reducing related costs, then a reasonable goal is to cut these costs in half in five years.
- *Do* have the figures reviewed in advance by those people in finance (and elsewhere) to whom upper management looks for checking the validity of financial figures.
- *Don't* inflate the present costs by including debatable or borderline items. The risk is that the decisive review meetings will get bogged down in debating the validity of the figures without ever discussing the merits of the proposals.

▲ *Don't* imply that the total costs will be reduced to zero. Any such implication will likewise divert attention from the merits of the proposals.
▲ *Don't* force the first few projects on managers who are not really sold on them or on unions that are strongly opposed. Instead, start in areas that show a climate of receptivity. The results obtained in these areas will determine whether the overall initiative will expand or die out.

The needs for breakthrough go beyond satisfying customers or making cost reductions. New forces keep coming over the horizon. Recent examples have included growth in product liability, the consumerism movement, foreign competition, legislation, and environmental concerns of all sorts. Breakthrough has provided much of the response to such forces.

Similarly, the means of convincing upper managers of the need for breakthrough go beyond reports from advocates. Conviction also may be supplied by visits to successful organizations, hearing papers presented at conferences, reading reports published by successful organizations, and listening to the experts, both internal and external. However, none of these is as persuasive as results achieved within one's own organization.

A final element of presentations to upper managers is to explain their personal responsibilities in launching and perpetuating breakthrough.

Mobilizing for Breakthrough

Until the 1980s, breakthrough in the West was not mandated—it was not a part of the business plan or a part of the job descriptions. Some breakthrough did take place, but on a voluntary basis. Here and there, a manager or a nonmanager, for whatever reason, elected to tackle some breakthrough project. He or she might persuade others to join an informal team. The result might be favorable, or it might not. This voluntary, informal approach yielded few breakthroughs. The emphasis remained on inspection, control, and firefighting.

The Need for Formality

The crisis that followed the Japanese revolution called for new strategies, one of which was a much higher rate of breakthrough. It then became evident that an informal approach would not produce thousands (or more) breakthroughs year after year. This led to experiments with structured approaches that in due course helped some organizations become role models.

Some upper managers protested the need for formality: "Why don't we just do it?" The answer depends on how many breakthroughs are needed. For just a few projects each year, informality is adequate; there is no need to mobilize. However, making breakthroughs by the hundreds or the thousands requires a formal structure.

As it has turns out, mobilizing for breakthrough requires two levels of activity, as shown in Table 6.3. One activity mobilizes the organization's resources to deal with the breakthrough projects collectively. This becomes the responsibility of management. The other activity is needed to carry out the projects individually. This becomes the responsibility of the breakthrough teams.

Table 6.3 Mobilizing for Breakthrough

Activities by management	Activities by project teams
Establish infrastructure: quality councils	Verify problem
Select problems; determine goals and targets	Analyze symptoms
Create project charters and assign teams	Theorize as to causes
Launch teams and review progress	Test theories
Provide recognition and rewards	Discover causes
	Stimulate remedies and controls

The Executive "Quality Council"

The first step in mobilizing for breakthrough is to establish the organization's council (or similar name). The basic responsibility of this council is to launch, coordinate, and "institutionalize" annual breakthrough. Such councils have been established in many organizations. Their experiences provide useful guidelines.

Membership and Responsibilities

Council membership is typically drawn from the ranks of senior managers. Often, the senior management committee is also the council. Experience has shown that councils are most effective when upper managers are personally the leaders and members of the senior councils.

In large organizations, it is common to establish councils at the divisional level as well as at the corporate level. In addition, some individual facilities may be so large as to warrant establishing a local council. When multiple councils are established, they are usually linked together—members of high-level councils serve as chairpersons of lower-level councils. Figure 6.3 is an example of such linkage.

Experience has shown that organizing councils solely in the lower levels of management is ineffective. Such organization limits breakthrough projects to the useful many while neglecting the vital few projects—those that can produce the greatest results. In addition, councils solely at lower levels send a message to all: "Breakthrough is not high on upper management's agenda."

It is important for each council to define and publish its responsibilities so that (1) the members agree on their goal and (2) the rest of the organization can become informed of upcoming events.

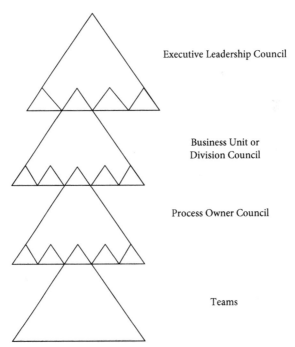

Figure 6.3 Quality councils are linked together. (From Juran Institute, Inc., 2013)

Many councils have published their statements of responsibility. Major common elements have included the following:

- Formulate the policies, such as focus on the customer has top priority, breakthrough must go on year after year, participation should be universal, or the reward system should reflect performance on breakthrough.
- Estimate the major dimensions, such as the status of the company's quality compared with its competitors', the extent of chronic waste, the adequacy of major business processes, or the results achieved by prior breakthroughs.
- Establish processes for selecting projects, such as soliciting and screening nominations, choosing projects to be tackled, preparing goal statements, or creating a favorable climate for breakthrough.
- Establish processes for carrying out the projects, such as selecting team leaders and members or defining the role of project teams.
- Provide support for the project teams, such as training time for working on projects, diagnostic support, facilitator support, or access to facilities for tests and tryouts.
- Establish measures of progress, such as effect on customer satisfaction, effect on financial performance, or extent of participation by teams.
- Review progress, assist teams in the event of obstacles, and ensure that remedies are implemented.
- Provide for public recognition of teams.
- Revise the reward system to reflect the changes demanded by introducing annual breakthrough.

Councils should anticipate the troublesome questions and, to the extent feasible, provide answers at the time of announcing the intention to go into annual breakthrough. Some senior managers have gone to the extent of creating a videotape to enable a wide audience to hear the identical message from a source of undoubted authority.

Leaders Must Face Up to the Apprehensions about Elimination of Jobs

Employees want not only dialogue on such an important issue, but also assurance relative to their apprehensions, notably the risk of job loss due to improvements. Most upper managers have been reluctant to face up to these apprehensions. Such reluctance is understandable. It is risky to provide assurances when the future is uncertain.

Nevertheless, some managers have estimated in some depth the two pertinent rates of change:

1. The rate of creation of job openings due to attrition: retirements, offers of early retirement, resignation, and so on. This rate can be estimated with a fair degree of accuracy.
2. The rate of elimination of jobs due to reduction of chronic waste. This estimate is more speculative—it is difficult to predict how soon the breakthrough rate will get up to speed. In practice, organizations have been overly optimistic in their estimates.

Analysis of these estimates can help managers judge what assurances they can provide, if any. It also can shed light on the choice of alternatives for action: retrain for jobs that have opened up, reassign to areas that have job openings, offer early retirement, assist in finding jobs in other organizations, and/or provide assistance in the event of termination.

Assistance from the Quality and/or Performance Excellence Functions

Many councils secure the assistance of the performance excellence and quality departments. These are specialists who are skilled in the methods and tools to attain high quality. They are there to:

- Provide inputs needed by the council for planning to introduce breakthrough
- Draft proposals and procedures
- Carry out essential details such as screening nominations for projects
- Develop training materials
- Develop new scorecards
- Prepare reports on progress

It is also usual for the quality directors to serve as secretaries of the council.

Breakthrough Goals in the Business Plan

Organizations that have become the market leaders—the role models—all adopted the practice of enlarging their business plan to include quality goals. In effect, they translated the threats and opportunities faced by their organizations into goals, such as:

- Increase on-time deliveries from 83 to 100 percent over the next two years.
- Reduce the cost of poor quality by 50 percent over the next five years.

Such goals are clear—each is quantified, and each has a timetable. Convincing upper managers to establish such goals is a big step, but it is only the first step.

Deployment of Goals

Goals are merely a wish list until they are deployed—until they are broken down into specific projects to be carried out and assigned to specific individuals or teams who are then provided with the resources needed to take action. Figure 6.4 shows the anatomy of the deployment process. In the figure, the broad (strategic) goals are established by the council and become a part of the organization's business plan. These goals are then divided and allocated to lower levels to be translated into action. In large organizations, there may be further subdivision before the action levels are reached. The final action level may consist of individuals or teams.

In response, the action levels select breakthrough projects that collectively will meet the goals. These projects are then proposed to the upper levels along with estimates of the resources needed. The proposals and estimates are discussed and revised until final decisions are reached. The end result is an agreement on which projects to tackle, what resources to provide, and who will be responsible for carrying out the projects.

This approach of starting at the top with strategic goals may seem like a purely top-down activity. However, the deployment process aims to provide open discussion in both directions before final decisions are made, and such is the way it usually works out.

The concept of strategic goals involves the vital few matters, but it is not limited to the corporate level. Goals also may be included in the busi-

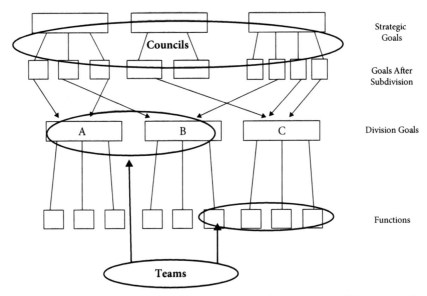

Figure 6.4 Anatomy of the deployment process. (From Juran and DeFeo, 2010)

ness plans of divisions, profit centers, field offices, and still other facilities. The deployment process is applicable to all these.

The Project Concept

As used here, a project is a chronic problem scheduled for solution. The project is the focus of actions for breakthrough. All breakthrough takes place project by project and in no other way.

Some projects are derived from the goals that are in the organization's business plan. These are relatively few in number, but each is quite important. Collectively, these are among the vital few projects (see "Use of the Pareto Principle"). However, most projects are derived not from the organization's business plan but from the nomination selection process, as discussed later.

Use of the Pareto Principle

A valuable aid to the selection of projects during the deployment process is the Pareto principle. This principle states that in any population that contributes to a common effect, a relative few of the contributors—the vital few—account for the bulk of the effect. The principle applies widely in human affairs. Relatively small percentages of the individuals write most of the books, commit most of the crimes, own most of the wealth, and so on.

Presentation of data in the form of a Pareto diagram greatly enhances communication of the information, most notably in convincing upper management of the source of a problem and gaining support for a proposed course of action to remedy the problem. (For an account of how Dr. Juran came to name the Pareto principle, see the Appendix in *Juran's Quality Handbook: The Complete Guide to Performance Excellence*, 6th ed.)

The Useful Many Problems and Solutions

Under the Pareto principle, the vital few projects provide the bulk of the breakthrough, so they receive top priority. Beyond the vital few are the useful many problems. Collectively, they contribute only a minority of the breakthrough, but they provide most of the opportunity for employee participation. The useful many projects are made through the application of workplace improvement teams, quality circles, the Lean 5S tools, or self-directed work teams.

The Nomination and Selection Process

Most projects are chosen through the nomination and selection process, involving several steps:

- Project nomination
- Project screening and selection
- Preparation and publication of project goal statements

Sources of Nominations

Nominations for projects can come from all levels of the organization. At the higher levels, the nominations tend to be extensive in size (the vital few) and multifunctional in scope. At lower levels, the nominations are smaller in size (the useful many) and tend to be limited in scope to the boundaries of a single department.

Nominations come from many sources. These include:

- *Formal data systems*, such as field reports on product performance, customer complaints, claims, returns, and so on; accounting reports on warranty charges and on internal costs of poor quality; and service call reports. (Some of these data systems provide for analyzing the data to identify problem areas.)
- *Special studies*, such as customer surveys, employee surveys, audits, assessments, benchmarking against competitors, and so on.
- *Reactions from customers* who have run into product dissatisfactions are often vocal and insistent. In contrast, customers who judge product features to be not competitive may simply (and quietly) become ex-customers.
- *Field intelligence* derived from visits to customers, suppliers, and others; actions taken by competitors; and stories published in the media (as reported by sales, customer service, technical service, and others).
- *The impact on society*, such as new legislation, extension of government regulation, and growth of product liability lawsuits.
- *The managerial hierarchy*, such as the council, managers, supervisors, professional specialists, and project teams.
- *The workforce* through informal ideas presented to supervisors, formal suggestions, ideas from circles, and so on.
- *Proposals* relating to business processes.

Criteria for Projects

During the beginning stages of project-by-project breakthrough, everyone is in a learning state. Projects are assigned to project teams, who are in training. Completing a project is a part of that training. Experience with such teams has evolved a broad set of criteria:

- The project should deal with a *chronic problem*—one that has been awaiting a solution for a long time.
- The project should be *feasible*. There should be a good likelihood of completing it within a few months. Feedback from organizations suggests that the most frequent reason for failure of the first project has been failure to meet the criterion of feasibility.

- ▲ The project should be *significant*. The end result should be sufficiently useful to merit attention and recognition.
- ▲ The results should be *measurable*, whether in money or in other significant terms.
- ▲ The first projects should be winners.

Additional criteria to select projects are aimed at what will do the organization the most good:

- ▲ *Return on investment*. This factor has great weight and is decisive, all other things being equal. Projects that do not lend themselves to computing return on investment must rely for their priority on managerial judgment.
- ▲ *The amount of potential breakthrough*. One large project will take priority over several small ones.
- ▲ *Urgency*. There may be a need to respond promptly to pressures associated with product safety, employee morale, and customer service.
- ▲ *Ease of technological solution*. Projects for which the technology is well developed will take precedence over projects that require research to discover the needed technology.
- ▲ *Health of the product line*. Projects involving thriving product lines will take precedence over projects involving obsolescent product lines.
- ▲ *Probable resistance to change*. Projects that will meet a favorable reception take precedence over projects that may meet strong resistance, such as from the labor union or from a manager set in his or her ways.

Most organizations use a systematic approach to evaluate nominations relative to these criteria. This yields a composite evaluation that then becomes an indication of the relative priorities of the nominations.

Project Selection

The result of the screening process is a list of recommended projects in their order of priority. Each recommendation is supported by the available information on compatibility with the criteria and potential benefits, resources required, and so on. This list is commonly limited to matters in which the council has a direct interest.

The council reviews the recommendations and makes the final determination on which projects are to be tackled. These projects then become an official part of the organization's business. Other recommended projects are outside the scope of the direct interest of the council. Such projects are recommended to appropriate subcouncils, managers, and so on. None of the preceding prevents projects from being undertaken at local levels by supervisors or by the workforce.

Vital Few and Useful Many

Some organizations completed many projects. Then, when questions were raised—"What have we gotten for all this effort?"—they were dismayed to learn that there was no noticeable effect on the bottom line. Investigation then showed that the reason could be traced to the process used for project selection. The projects actually selected had consisted of:

- *Firefighting projects.* These are special projects for getting rid of sporadic "spikes." Such projects did not attack the chronic waste and hence could not improve financial performance.
- *Useful many projects.* By definition, these have only a minor effect on financial performance but have great effect on human relations.
- *Projects for improving human relations.* These can be quite effective in their field, but the financial results are usually not measurable.

To achieve a significant effect on the bottom line requires selecting the vital few projects as well as the useful many. It is feasible to work on both, since different people are assigned to each.

There is a school of thought emerging that contends that the key to market leadership is "tiny breakthroughs in a thousand places"—in other words, the useful many. Another school urges focus on the vital few. In Dr. Juran's experience, neither of these schools has the complete answer. Both are needed—at the right time.

The vital few projects are the major contributors to leadership and to the bottom line. The useful many projects are the major contributors to employee participation and to the quality of work life. Each is necessary; neither is sufficient.

The vital few and useful many projects can be carried out simultaneously. Successful organizations have done just that by recognizing that while there are these two types of projects, they require the time of different categories of organization personnel.

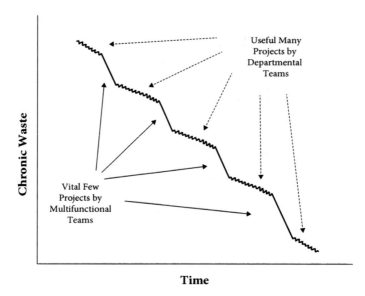

Figure 6.5 Interrelation of projects. (From Juran Institute, Inc., 2013)

The interrelation of these two types of projects is shown in Figure 6.5, where the horizontal scale is time. The vertical scale is chronic waste. What goes up is bad. The useful many breakthroughs collectively create a gradually sloping line. The vital few breakthroughs, though less frequent, contribute the bulk of the total breakthrough.

Cost Figures for Projects

To meet the preceding criteria (especially that of return on investment) requires information on various costs:

- The cost of chronic waste associated with a given nomination
- The potential cost reduction if the project is successful
- The cost of the needed diagnosis and remedy

Costs versus Percentage of Deficiencies

It is risky to judge priorities based solely on the percentage of deficiencies (errors, defects, and so on). On the face of it, when this percentage is low, the priority of the nomination also should be low. In some cases this is true, but in others it can be seriously misleading.

Elephant-Sized and Bite-Sized Projects

There is only one way to eat an elephant: bite by bite. Some projects are "elephant-sized"; that is, they cover so broad an area of activity that they must be subdivided into multiple "bite-sized" projects. In such cases, one project team can be assigned to "cut up the elephant." Other teams are then assigned to tackle the resulting bite-sized projects. This approach shortens the time to complete the project, since the teams work concurrently. In contrast, use of a single team stretches the time out to several years. Frustration sets in, team membership changes due to attrition, the project drags, and morale declines.

A most useful tool for cutting up the elephant is the Pareto analysis. For elephant-sized projects, separate goal statements are prepared for the broad coordinating team and for each team assigned to a bite-sized project.

Replication and Cloning

Some organizations consist of multiple autonomous units that exhibit much commonality. A widespread example is the chains of retail stores, repair shops, hospitals, and so on. In such organizations, a breakthrough project that is carried out successfully in one operating unit logically becomes a nomination for application to other units. This is called *cloning* the project.

It is quite common for the other units to resist applying the breakthrough to their operation. Some of this resistance is cultural in nature (not invented here, and so on). Other resistance may be due to real differences in operating conditions. For example, telephone exchanges perform similar functions for their customers. However, some serve mainly industrial customers, whereas others serve mainly residential customers.

Upper managers are wary of ordering autonomous units to clone breakthroughs that originated elsewhere. Yet cloning has advantages. Where feasible, it provides additional breakthroughs without the need to duplicate the prior work of diagnosis and design of remedy.

What has emerged is a process as follows:

- Project teams are asked to include in their final report their suggestions as to sites that may be opportunities for cloning.
- Copies of such final reports go to those sites.
- The decision of whether to clone is made by the sites.

However, the sites are required to make a response as to their disposition of the matter. This response is typically in one of three forms:

- We have adopted the breakthrough.
- We will adopt the breakthrough, but we must first adapt it to our conditions.
- We are not able to adopt the breakthrough for the following reasons.

In effect, this process requires the units to adopt the breakthrough or give reasons for not doing so. The units cannot just quietly ignore the recommendation.

A more subtle but familiar form of cloning is done through projects that have repetitive application over a wide variety of subject matter.

A project team develops computer software to find errors in spelling. Another team evolves an improved procedure for processing customer orders through the organization. A third team works up a procedure for conducting design reviews. What is common about such projects is that the result permits repetitive application of the same process to a wide variety of subject matter: many different misspelled words, many different customer orders, and many different designs.

Model of the Infrastructure

There are several ways to show in graphical form the infrastructure for breakthrough—the elements of the organization, how they relate to one another, and the flow of events. Figure 6.6 shows the elements of the infrastructure in pyramid form. The pyramid depicts a hierarchy consisting of top management, the autonomous operating units, and the major staff functions. At the top of the pyramid are the corporate council and the subsidiary councils, if any. Below these levels are the multifunctional breakthrough teams. (There may be a committee structure between the councils and the teams.)

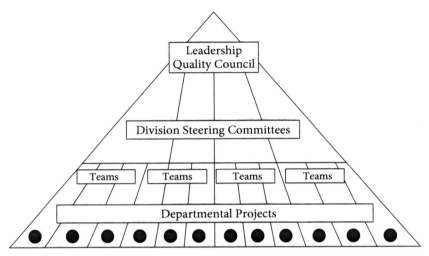

Figure 6.6 Model of the infrastructure for breakthrough quality improvement. (From Juran and DeFeo, 2010)

At the intradepartmental level are teams from the workforce—circles or other forms. This infrastructure permits employees in all levels of the organization to participate in breakthrough projects, the useful many as well as the vital few.

Team Organization

Breakthrough teams do not appear on the organization chart. Each "floats"—it has no personal boss. Instead, the team is supervised *impersonally* by its goal statement and by the breakthrough roadmap.

The team does have its own internal organizational structure. This structure invariably includes a team *leader* (chairperson and so on) and a team *secretary*. In addition, there is usually a *facilitator*.

The Team Leader

The leader is usually appointed by the sponsor—the council or other supervising group. Alternatively, the team may be authorized to elect its leader.

The leader has several responsibilities. As a team member, the leader *shares* in the responsibility for completing the team's goal. In addition, the leader has administrative duties. These are *unshared* and include:

- Ensuring that meetings start and finish on time
- Helping the members attend the team meetings
- Ensuring that the agendas, minutes, reports, and so on are prepared and published
- Maintaining contact with the sponsoring body

Finally, the leader has the responsibility of *oversight*. This is met not through the power of command—the leader is not the boss of the team. It is met through the power of leadership. The responsibilities include:

- Orchestrating the team activities
- Stimulating all members to contribute
- Helping to resolve conflicts among members
- Assigning the homework to be done between meetings

To meet such responsibilities requires multiple skills, which include:

- A trained capability for leading people
- Familiarity with the subject matter of the goal
- A firm grasp of the breakthrough process and the associated tools

The Team Members

Team members as used here include the team leader and secretary. The responsibilities of any team member consist mainly of the following:

- Arranging to attend the team meetings
- Representing his or her department
- Contributing job knowledge and expertise
- Proposing theories of causes and ideas for remedies
- Constructively challenging the theories and ideas of other team members
- Volunteering for or accepting assignments for homework

Finding the Time to Work on Projects

Work on project teams is time consuming. Assigning someone to a project team adds about 10 percent to that person's workload. This added time is needed to attend team meetings, perform the assigned homework, and so on. Finding the time to do all this is a problem to be solved, since this added work is thrust on people who are already fully occupied.

No upper manager known to Dr. Juran has been willing to solve the problem by hiring new people to make up for the time demanded by the breakthrough projects. Instead, it has been left to each team member to solve the problem in her or his own way. In turn, the team members have adopted such strategies as:

- Delegating more activities to subordinates
- Slowing down the work on lower-priority activities
- Improving time management on the traditional responsibilities
- Looking for ongoing activities that can be terminated (In several organizations, there has been a specific drive to clear out unneeded work to provide time for breakthrough projects.)

As projects begin to demonstrate high returns on investment, the climate changes. Upper managers become more receptive to providing resources. In addition, the successful projects begin to reduce workloads that previously were inflated by the presence of chronic wastes.

Facilitators and Black Belts

Most organizations make use of internal consultants, usually called *facilitators* or *Black Belts*, to assist teams. A facilitator like a Black Belt does not have to be a member of the team and may not have any responsibility for carrying out the team goal. (The literal meaning of the word *facilitate* is "to make things easy.") The prime role of the facilitator is to help the team to carry out its goal. The usual roles of facilitators consist of a selection from the following:

- *Explain the organization's intentions.* The facilitator usually has attended briefing sessions that explain what the organization is trying to accomplish. Much of this briefing is of interest to the project teams.
- *Assist in team building.* The facilitator helps the team members learn to contribute to the team effort: propose theories, challenge the theories of others, and/or propose lines of investigation. Where the team concept is new to an organization, this role may require working directly with individuals to stimulate those who are unsure about how to con-

tribute and to restrain the overenthusiastic ones. The facilitator also may evaluate the progress in team building and provide feedback to the team.

- *Assist in training.* Most facilitators have undergone training in team building and in the breakthrough process. They usually have served as facilitators for other teams. Such experiences qualify them to help train project teams in several areas: team building, the breakthrough roadmap, and/or use of the tools.
- *Relate experiences from other projects.* Facilitators have multiple sources of such experiences:
    - Project teams previously served on
    - Meetings with other facilitators to share experiences in facilitating project teams
    - Final published reports of project teams
    - Projects reported in the literature
- *Assist in redirecting the project.* The facilitator maintains a detached view that helps him or her sense when the team is getting bogged down. As the team gets into the project, it may find itself getting deeper and deeper into a swamp. The project goal may turn out to be too broad, vaguely defined, or not doable. The facilitator usually can sense such situations earlier than the team and can help guide it to a redirection of the project.
- *Assist the team leader.* Facilitators provide such assistance in various ways:
    - Assist in planning the team meetings. This may be done with the team leader before each meeting.
    - Stimulate attendance. Most nonattendance is due to conflicting demands made on a team member's time. The remedy often must come from the member's boss.
    - Improve human relations. Some teams include members who have not been on good terms with one another or who develop friction as the project moves along. As an "outsider," the facilitator can help to direct the energies of such members into constructive channels. Such action usually takes place outside the team meetings. (Sometimes the leader is part of the problem. In such cases, the facilitator may be in the best position to help.)

- ▼ Assist on matters outside the team's sphere of activity. Projects sometimes require decisions or actions from sources that are outside the easy reach of the team. Facilitators may be helpful due to their wider range of contacts.
- ▲ *Support the team members.* Such support is provided in multiple ways:
  - ▼ Keep the team focused on the goal by raising questions when the focus drifts.
  - ▼ Challenge opinionated assertions by questions such as, Are there facts to support that theory?
  - ▼ Provide feedback to the team based on perceptions from seeing the team in action.
- ▲ *Report progress to the councils.* In this role, the facilitator is a part of the process of reporting on progress of the projects collectively. Each project team issues minutes of its meetings. In due course, each also issues its final report, often including an oral presentation to the council.

However, reports on the projects collectively require an added process. The facilitators are often a part of this added reporting network.

The Qualifications of Facilitators and Black Belts

Facilitators undergo special training to qualify them for these roles. The training includes skills in team building, resolving conflicts, communication, and management of quality change; knowledge relative to the breakthrough processes, for example, the breakthrough roadmap and the tools and techniques; and knowledge of the relationship of breakthrough to the organization's policies and goals. In addition, facilitators acquire maturity through having served on project teams and providing facilitation to teams.

This prerequisite training and experience are essential assets to the facilitator. Without them, she or he has great difficulty winning the respect and confidence of the project's team.

Most organizations are aware that to go into a high rate of breakthrough requires extensive facilitation. In turn, this requires a buildup of trained facilitators. However, facilitation is needed mainly during the start-up phase. Then, as team leaders and members acquire training and experience, there is less need for facilitator support. The buildup job becomes a maintenance job.

This phased rise and decline has caused most organizations to avoid creating full-time facilitators or a facilitator career concept. Facilitation is done on a part-time basis. Facilitators spend most of their time on their regular job.

In many larger organizations, Black Belts are full-time specialists. Following intensive training in the breakthrough process, these persons devote all their time to the breakthrough activity. Their responsibilities go beyond facilitating project teams and may include:

- Assisting in project nomination and screening
- Conducting training courses in the methods and tools
- Coordinating the activities of the project team with other activities in the organization, including conducting difficult analyses
- Assisting in the preparation of summarized reports for upper managers

A team has no personal boss. Instead, the team is supervised impersonally. Its responsibilities are defined in the following:

- *The project charter.* This goal statement is unique to each team.
- *The steps or universal sequence for breakthrough.* This is identical for all teams. It defines the actions to be taken by the team to accomplish its goal.

The project team has the principal responsibility for the steps that now follow—taking the two "journeys." The diagnostic and remedial journeys are as follows:

- *The diagnostic journey from symptom to cause.* It includes analyzing the symptoms, theorizing as to the causes, testing the theories, and establishing the causes.
- *The remedial journey from cause to remedy.* It includes developing the remedies, testing and proving the remedies under operating conditions, dealing with resistance to change, and establishing controls to hold the gains.

Diagnosis is based on the factual approach and requires a firm grasp of the meanings of key words. It is helpful to define some of these key words at the outset.

Leaders Must Learn Key Breakthrough Terminology

A *defect* is any state of unfitness for use or nonconformance to specification. Examples are illegible invoices, scrap, and low mean time between failures. Other names include *error, discrepancy,* and *nonconformance.*

A *symptom* is the outward evidence that something is wrong or that there is a defect. A defect may have multiple symptoms. The same word may serve as a description of both defect and symptom.

A *theory* or *hypothesis* is an unproved assertion as to reasons for the existence of defects and symptoms. Usually, multiple theories are advanced to explain the presence of defects.

A *cause* is a proved reason for the existence of a defect. Often there are multiple causes, in which case they follow the Pareto principle—the vital few causes will dominate all the rest.

A *dominant cause* is a major contributor to the existence of defects and one that must be remedied before there can be an adequate breakthrough.

Diagnosis is the process of studying symptoms, theorizing as to causes, testing theories, and discovering causes.

A *remedy* is a change that can eliminate or neutralize a cause of defects.

Diagnosis Should Precede Remedy

It may seem obvious that diagnosis should precede remedy, yet biases or outdated beliefs can get in the way.

For example, during the twentieth century, many upper managers held deep-seated beliefs that most defects were due to workforce errors. The facts seldom bore this out, but the belief persisted. As a result, during the 1980s, many of these managers tried to solve their problems by exhorting the workforce to make no defects. (In fact, defects are generally over 80 percent management-controllable and under 20 percent worker-controllable.)

Untrained teams often try to apply remedies before the causes are known. ("Ready, fire, aim.") For example:

- ▲ An insistent team member "knows" the cause and pressures the team to apply a remedy for that cause.
- ▲ The team is briefed on the technology by an acknowledged expert. The expert has a firm opinion about the cause of the symptom, and the team does not question the expert's opinion.

- As team members acquire experience, they also acquire confidence in their diagnostic skills. This confidence then enables them to challenge unproved assertions.
- Where deep-seated beliefs are widespread, special research may be needed.

Institutionalizing Breakthrough

Numerous organizations have initiated breakthrough, but few have succeeded in institutionalizing it so that it goes on year after year. Yet many of these organizations have a long history of annually conducting product development, cost reduction, productivity breakthrough, and so on. The methods they used to achieve such annual breakthrough are well known and can be applied to breakthrough. They are as follows:

- Enlarge the annual business plan to include goals for breakthrough.
- Make breakthrough a part of everyone's job description. In most organizations, the activity of breakthrough has been regarded as incidental to the regular job of meeting the goals for cost, delivery, and so on. The need is to make breakthrough a part of the regular job.
- Establish upper management audits that include review of progress on breakthrough.
- Revise the merit rating and reward system to include a new parameter—performance on breakthrough—and give it proper weight.
- Create well-publicized occasions to provide recognition for performance on breakthrough.

Review Progress

Scheduled, periodic reviews of progress by upper managers are an essential part of maintaining annual breakthroughs. Activities that do not receive such review cannot compete for priority with activities that do receive such review. Subordinates understandably give top priority to matters that are reviewed regularly by their superiors.

There is also a need for regular review of the breakthrough process. This is done through audits that may extend to all aspects of managing for quality. Much of the database for progress review comes from the reports issued by the project teams. However, it takes added work to analyze these reports and to prepare the summaries needed by upper managers. Usually, this added work is done by the secretary of the council with the aid of the facilitators, the team leaders, and other sources such as finance.

As organizations gain experience, they design standardized reporting formats to make it easy to summarize reports by groups of projects, by product lines, by business units, by divisions, and for the corporation. One such format, used by a large European organization, determines the following for each project:

- The original estimated amount of chronic waste.
- The original estimated reduction in cost if the project were to be successful.
- The actual cost reduction achieved.
- The capital investment.
- The net cost reduction.
- The summaries are reviewed at various levels. The corporate summary is reviewed quarterly at the chairperson's staff meeting (personal communication to the author).

References

Juran, J. M. (1964). *Managerial Breakthrough*. McGraw-Hill, New York. Revised edition, 1995.

Juran, J. M. (1985). "A Prescription for the West—Four Years Later." European Organization for Quality, 29th Annual Conference. Reprinted in *The Juran Report*, no. 5, Summer 1985.

Juran, J. M. (1988). "Juran on Quality Leadership," a video package, Juran Institute, Inc., Wilton, Conn.

Juran, J. M. and DeFeo, J. A. (2010). *Juran's Quality Handbook: The Complete Guide to Performance Excellence*, 6th ed. McGraw-Hill, New York.

Sakai, S. (1994). "Rediscovering Quality—The Toyota Way." IMPRO 1994 Conference Proceedings, Juran Institute, Inc., Wilton, Conn.

Welch, J. (2005). *Winning*. HarperCollins, New York.

CHAPTER 7

Assuring Repeatable and Compliant Processes

This chapter describes the compliance process or simply the *control process*. *Control* is a universal managerial process to ensure that all key operational processes are stable—to prevent adverse change and to "ensure that the planned performance targets are met." Control includes product control, service control, process control, and even facilities control. To maintain stability, the control process evaluates actual performance, compares actual performance to targets, and takes action on any differences.

Compliance and Control Defined

Compliance or *quality control* is the third universal process in the Juran Trilogy. The others are quality planning and quality improvement. The Juran Trilogy diagram (Figure 7.1) shows the interrelationship of these processes.

Figure 7.1 is used in several other areas in this book to describe the relationships among planning, improvement, and control—the fundamental managerial processes in quality management. What is important for this chapter is to concentrate on the two "zones of control."

In Figure 7.1, we can easily see that although the process is in control in the middle of the chart, we are running the process at an unacceptable level of performance and "waste." What is necessary here is not more control, but improvement—actions to change the level of performance.

After the improvements have been made, a new level of performance has been achieved. Now it is important to establish new controls at this level to prevent the performance level from deteriorating to the previous level or even worse. This is indicated by the second zone of control.

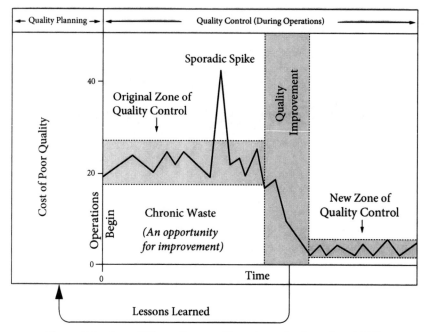

Figure 7.1 Juran Trilogy diagram. (From Juran Institute, Inc., 2013)

The term *control of quality* emerged early in the twentieth century (Radford, 1917, 1922). The concept was to broaden the approach to achieving quality, from the then-prevailing after-the-fact inspection (detection control) to what we now call *prevention* (*proactive control*). For a few decades, the word *control* had a broad meaning, which included the concept of quality planning. Then came events that narrowed the meaning of quality control. The *statistical quality control* movement gave the impression that quality control consisted of using statistical methods. The *reliability* movement claimed that quality control applied only to quality at the time of test but not during service life.

In the United States, the term *quality control* now often has the meaning defined previously. It is a piece of a "performance excellence, operational excellence, business excellence, or total quality program," which are now used interchangeably to comprise the all-embracing term to describe the methods, tools, and techniques to manage the quality of an organization.

In Japan, the term *quality control* retains a broad meaning. Their *total quality control* is equivalent to our term *business excellence*. In 1997, the Union of Japanese Scientists and Engineers (JUSE) adopted the term *Total*

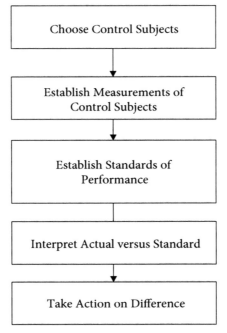

Figure 7.2 Input-output diagram.

Quality Management (TQM) to replace *Total Quality Control* (TQC) to more closely align themselves with the more common terminology used in the rest of the world.

Figure 7.2 shows the input-output features of this step.

The Relation to Quality Assurance

Quality control and quality assurance have much in common. Each evaluates performance. Each compares performance to goals. Each acts on the difference. However, they also differ from each other. Quality control is, as its primary purpose, maintaining control. Performance is evaluated during operations, and performance is compared to targets during operations. In the process, metrics are utilized to monitor adherence to standards. The resulting information is received and used by the employees.

The main purpose of quality assurance is to verify that control is being maintained. Performance is evaluated after operations, and the resulting information is provided to both the employees and others who have a need to know. Results metrics are utilized to determine conformance to customer needs and expectations. Others may include leadership; plant; functional; corporate staffs; regulatory bodies; customers; and the general public.

The Feedback Loop

Quality control takes place by use of the feedback loop. A generic form of the feedback loop is shown in Figure 7.3.

The progression of steps in Figure 7.3 is as follows:

1. A sensor is "plugged in" to evaluate the actual quality of the control subject—the product or process feature in question. The performance of a process may be determined directly by evaluation of the process feature, or indirectly by evaluation of the product feature—the product "tells" on the process.
2. The sensor reports the performance to an umpire.
3. The umpire also receives information on the quality goal or standard.
4. The umpire compares actual performance to standard. If the difference is too great, the umpire energizes an actuator.
5. The actuator stimulates the process (whether human or technological) to change the performance so as to bring quality into line with the quality goal.
6. The process responds by restoring conformance.

Note that in Figure 7.3 the elements of the feedback loop are functions. These functions are universal for all applications, but responsibility for carrying out these functions can vary widely. Much control is carried

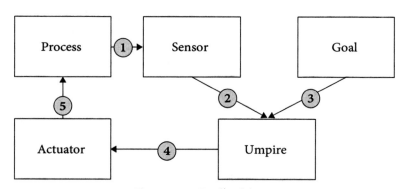

Figure 7.3 Feedback loop.

out through automated feedback loops. No human beings are involved. Common examples are the thermostat used to control temperature and the cruise control used in automobiles to control speed.

Another form of control is self-control carried out by employees. An example of such self-control is the village artisan who performs every one of the steps of the feedback loop. The artisan chooses the control subjects based on understanding the needs of customers, sets the quality targets to meet the needs, senses the actual quality performance, judges conformance, and becomes the actuator in the event of nonconformance.

This concept of self-control is illustrated in Figure 7.4. The essential elements here are the need for the employee or work team to know what they are expected to do, to know how they are actually doing, and to have the means to regulate performance. This implies that they have a capable process and have the tools, skills, and knowledge necessary to make the adjustments and the authority to do so.

A further common form of feedback loop involves office clerks or factory workers whose work is reviewed by umpires in the form of inspectors. This design of a feedback loop is largely the result of the Taylor management system adopted in the early twentieth century. It focused on the separation of planning for quality from the execution or operations. The Taylor management system emerged a century ago and contributed greatly to increasing productivity. However, the effect on quality was largely negative. The negative impact resulted in large costs associated with poor quality, products and services that have higher levels of failure, and customer dissatisfaction.

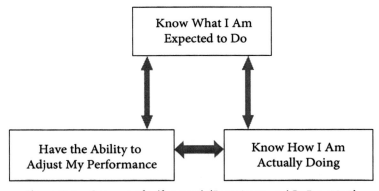

Figure 7.4 Concept of self-control. (From Juran and DeFeo, 2010)

The Elements of the Feedback Loop

The feedback loop is universal. It is fundamental to maintaining control of every process. It applies to all types of operations, whether in service industries or manufacturing industries, whether for profit or not. The feedback loop applies to all levels in the hierarchy, from the chief executive officer to the members of the workforce. However, there is wide variation in the nature of the elements of the feedback loop.

In Figure 7.5 a simple flowchart is shown that describes the control process with the simple universal feedback loop embedded.

The Control Subjects

Each feature of the product (goods and services) or process becomes a control subject (the specific attribute or variable to be controlled)—a center around which the feedback loop is built. The critical first step is to choose the control subject. To choose control subjects, you should identify the major work processes and products; define the objectives of the work pro-

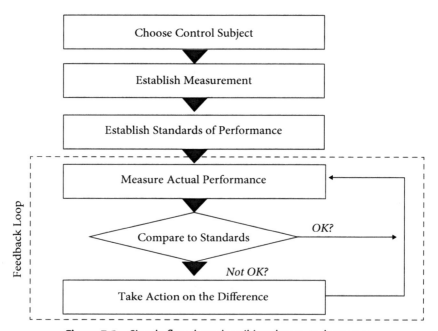

Figure 7.5 Simple flowchart describing the control process.

cesses; succinctly define the work processes; identify the customers of the process, and then select the control subjects (Key Product Characteristics (KPCs) and/or Key Control Characteristics (KCCs)). Control subjects are derived from multiple sources, which include:

- Stated customer needs for product features
- Translated "voice of the customer" needs into product features
- Defined process features that create the product or service features
- Industry and government standards and regulations (i.e., Sarbanes-Oxley, ISO 9000, etc.)
- Need to protect human safety and the environment (i.e., OSHA, ISO 14000)
- Need to avoid side effects such as irritations to stakeholders, employees, or a neighboring community
- Failure mode and effects analyses
- Control plans
- Results of design of experiments

At the staff level, control subjects consist mainly of product and process features defined in technical specifications and procedures manuals. At managerial levels, the control subjects are broader and increasingly business-oriented. Emphasis shifts to customer needs and to competition in the marketplace. This shift in emphasis then demands broader control subjects, which, in turn, have an influence on the remaining steps of the feedback loop.

Establish Measurement

After the control subjects are chosen, the next step is to establish the means of measuring the actual performance of the process or the quality level of the goods or services being created. Measurement is one of the most difficult tasks of management. In establishing the measurement, we need to clearly specify the means of measuring (the sensor), the accuracy and precision of the measurement tool, the unit of measure, the frequency of measuring, the means by which data will be recorded, the format for reporting the data, the analysis to be made on the data to convert them to usable information, and who will make the measurement. In establishing the unit of measure, one should select a unit of measure that is understandable, provides an agreed-upon basis for decision making, is customer-focused, and can be applied broadly.

Establish Standards of Performance: Product Goals and Process Goals

For each control subject it is necessary to establish a standard of performance—a target or goal (also metrics, objectives, etc.). A standard of performance is an aimed-at target toward which work is expended. Table 7.1 gives some examples of control subjects and the associated goals.

Table 7.1 Control Subjects and Associated Quality Goals

Control Subject	Goal
Vehicle mileage	Minimum of 25 mi/gal highway driving
Overnight delivery	99.5% delivered prior to 10:30 a.m. next morning
Reliability	Fewer than 3 failures in 25 years of service
Temperature	Minimum 505°F; maximum 515°F
Purchase-order error rate	No more than 3 errors/1000 purchase orders
Competitive performance	Equal to or better than top three competitors on six factors
Customer satisfaction	90% or better rate, service outstanding or excellent
Customer retention	95% retention of key customers from year to year
Customer loyalty	100% of market share of over 80% of customers

The prime goal for products and services is to meet customer needs. Industrial customers often specify their needs with some degree of precision. Such specified needs then become goals for the producing company. In contrast, consumers tend to state their needs in vague terms. Such statements must then be translated into the language of the producer in order to become product goals.

Other goals for products that are also important are those for reliability and durability. Whether the products and services meet these goals can have a critical impact on customer satisfaction, loyalty, and overall costs. The failures of products under warranty can seriously affect the profitability of a company through both direct and indirect costs (loss of repeat sales, word of mouth, etc.).

The processes that produce products have two sets of goals:

1. To produce products and services that meet customer needs. Ideally, each and every unit produced should meet customer needs (meet specifications).
2. To operate in a stable and predictable manner. In the dialect of the quality specialist, each process should be "in a state of control." We will later elaborate on this in the section "Process Conformance."

Quality targets may also be established for functions, departments, or people. Performance against such goals then becomes an input to the company's scorecard, dashboard, and reward system. Ideally, such goals should be:

- *Legitimate.* They should have undoubted official status.
- *Measurable.* They can be communicated with precision.
- *Attainable.* This is evidenced by the fact that they have already been attained by others.
- *Equitable.* Attainability should be reasonably alike for individuals with comparable responsibilities.

Quality goals may be set from a combination of the following bases:

- Goals for product and service features and process features are largely based on technological analysis.
- Goals for functions, departments, and people should be based on the need of the business and external benchmarking rather than historical performance.

In the later 2000s, quality goals used at the highest levels of an organization have become commonplace. Establishing long-term goals such as reducing the costs of poor quality or becoming best in class have become a normal part of strategic business plans. The emerging practice is to establish goals on "metrics that matter," such as meeting customers' needs, exceeding the competition, maintaining a high pace of improvement, improving the effectiveness of business processes, and setting stretch goals to avoid failure-prone products and processes.

Measure Actual Performance

A critical step in controlling quality characteristics is to measure the actual performance of a process as precisely as possible. To do this requires measuring with a *sensor*. A sensor is a device or a person that makes the actual measurement.

The Sensor

A sensor is a specialized detecting device. It is designed to recognize the presence and intensity of certain phenomena and to convert the resulting data into *information*. This information then becomes the basis of decision making. At the lower levels of an organization, the information is often on a real-time basis and is used for daily control. At higher levels, the information is summarized in various ways to provide broader measures, detect trends, and identify the vital few problems.

The wide variety of control subjects requires a wide variety of sensors. A major category is the numerous technological instruments used to measure product features and process features. Familiar examples are thermometers, clocks, and weight scales. Another major category of sensors is the data systems and associated reports, which supply summarized information to the managerial hierarchy. Yet another category involves the use of human beings as sensors. Questionnaires, surveys, focus groups, and interviews are also forms of sensors.

Sensing for control is done on an organization level. Information is needed to manage for the short and long term. This has led to the use of computers to aid in the sensing and in converting the resulting data into information.

Most sensors provide their evaluations in terms of a unit of measure—a defined amount of some feature—which permits evaluation of that feature in numbers or pictures. Familiar examples of units of measure are degrees of temperature, hours, inches, and tons. A considerable amount of sensing is done by human beings. Such sensing is subject to numerous sources of error. The use of pictures as a standard of comparison can help reduce human errors. Also of vital importance to alleviate human errors is the application of detailed instructions.

Compare to Standards

The act of comparing to standards is often seen as the role of an umpire. The umpire may be a person or a technological device. Either way, the umpire may be called on to carry out any of or all the following activities:

- Compare the actual process performance to the targets.
- Interpret the observed difference (if any); determine if there is conformance to the target.
- Decide on the action to be taken.
- Stimulate corrective action.
- Record the results.

These activities require elaboration and will be examined more closely in an upcoming section.

Take Action on the Difference

In any well-functioning control system, we need a means of taking action on any difference between desired standards of performance and actual performance. For this we need an actuator. This device (human or technological or both) is the means for stimulating action to restore conformance. At the operations or employee level, it may be a keypad for giving orders to a centralized computer database, a change in a new procedure, a new specification document, or a new setting of a dial to adjust a machine to the right measure. At the management level, it may be a memorandum to subordinates, a new company policy, or a team to change a process.

The Key Process

In the preceding discussion we have assumed a process. This may also be human or technological or both. It is the means for producing the product and service features, each of which requires control subjects to ensure conformance to specifications. All work is done by a process. A process consists of inputs, labor, technology, procedures, energy, materials, and outputs.

Taking Corrective Action

There are many ways of taking corrective action to troubleshoot a process and return to the status quo. A popular example of a root cause and corrective action method is the so-called PDCA or PDSA cycle (first popularized by Walter Shewhart and then by Dr. Deming as the Deming wheel), as shown in Figure 7.6. Deming (1986) referred to this as the *Shewhart cycle*, which is the name many still use when describing this version of the feedback loop.

In this example, the feedback loop is divided into four steps labeled *plan, do, check, act* (PDCA) or *plan, do, study, act* (PDSA). This model is used by many health care and service industries. These steps correspond roughly to the following:

- *Plan* includes choosing control subjects and setting goals.
- *Do* includes running and monitoring the process.
- *Check* or *study* includes sensing and umpiring.
- *Act* includes stimulating the actuator and taking corrective action.

An early version of the PDCA cycle was included in W. Edwards Deming's first lectures in Japan (Deming, 1950). Since then, additional ver-

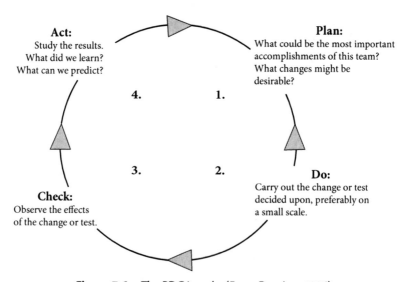

Figure 7.6 The PDCA cycle. (From Deming, 1986)

sions have been used, such as PDSA, PDCA, Root Cause Corrective Action (RCCA), and so on.

Some of these versions have attempted to label the PDCA cycle in ways that make it serve as a universal series of steps for both control and improvement. The authors feel that this confuses matters, since two very different processes are involved. Our experience is that all organizations should define two separate methods. One is to take corrective action on a *sporadic change* in performance.

RCCA, PDSA, and PDCA differ from improvement methods like Six Sigma in that the scope of the problem lends itself to a simpler, less complex analysis to find the root cause of a sporadic problem. RCCA analytical and communication tools contribute to the reduction of day-to-day problems that plague processes. Tools utilized for analysis and diagnosis of sporadic spikes typically take the form of graphical tools with less emphasis on statistical applications. Often many organizations that have been trained in RCCA and the like do not have the right tools and methods to solve chronic problems. It is best to use the Six Sigma D-M-A-I-C (Design – Measure – Analyze – Improve – Control) improvement methods.

The Pyramid of Control

Control subjects run to large numbers, but the number of "things" to be controlled is far larger. These things include the published catalogs and price lists sent out, multiplied by the number of items in each; the sales made, multiplied by the number of items in each sale; the units of product produced, multiplied by the associated numbers of quality features; and so on for the numbers of items associated with employee relations, supplier relations, cost control, inventory control, product and process developments, etc.

A study in one small company employing about 350 people found that there were more than a billion things to be controlled (Juran, 1964, pp. 181–182).

There is no possibility for upper leaders to control huge numbers of control subjects. Instead, they divide up the work of control using a plan of delegation similar to that shown in Figure 7.7.

This division of work establishes three areas of responsibility for control: control by nonhuman means, control by the workforce, and control by the managerial hierarchy.

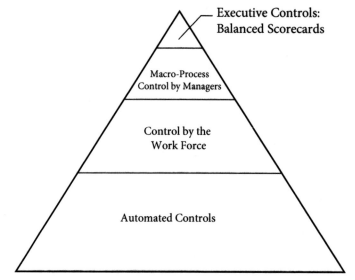

Figure 7.7 The pyramid of control. (From "Making Quality Happen," Juran Institute, Inc., Senior Executive Workshop, 1988, p. F-5)

Control by Technology (Nonhuman Means)

At the base of the pyramid are the automated feedback loops and error-proofed processes, which operate with no human intervention other than maintenance of facilities (which, however, is critical). These nonhuman methods provide control over a great majority of things. The control subjects are exclusively technological, and control takes place on a real-time basis.

The remaining controls in the pyramid require human intervention. By a wide margin, the most amazing achievement in quality control takes place during a biological process that is millions of years old—the growth of the fertilized egg into an animal organism. In human beings the genetic instructions that program this growth consist of a sequence of about 3 billion "letters." This sequence—the human genome—is contained in two strands of DNA (the double helix), which "unzip" and replicate billions of times during the growth process from fertilized egg to birth of the human being.

Given such huge numbers, the opportunities for error are enormous. (Some errors are harmless, but others are damaging and even lethal.) Yet the actual error rate is of the order of about 1 in 10 billion. This incredibly low error rate is achieved through a feedback loop involving three processes (Radman and Wagner, 1988):

- A high-fidelity selection process for attaching the right "letters," using chemical lock-and-key combinations
- A proofreading process for reading the most recent letter, and removing it if incorrect
- A corrective action process to rectify the errors that are detected

Control by the Employees (Workforce)

Delegating such decisions to the workforce yields important benefits in human relations and in conduct of operations. These benefits include shortening the feedback loop; providing the workforce with a greater sense of ownership of the operating processes, often referred to as *empowerment*; and liberating supervisors and leaders to devote more of their time to planning and improvement.

It is feasible to delegate many quality control decisions to the workforce. Many organizations already do. However, to delegate process control decisions requires meeting the criteria of *self-control* or *self-management*.

Control by the Managerial Hierarchy

The peak of the pyramid of control consists of the vital few control subjects. These are delegated to the various levels in the managerial hierarchy, including the upper leaders.

Leaders should avoid getting too deeply into making decisions on quality control. Instead, they should:

- Make the vital few decisions.
- Provide criteria to distinguish the vital few decisions from the rest. For an example of providing such criteria see Table 7.3 later in this chapter.
- Delegate the rest under a decision-making process that provides the essential tools and training.

The distinction between vital few matters and others originates with the control subjects. Table 7.2 shows how control subjects at two levels—workforce and upper management—affect the elements of the feedback loop.

Table 7.2 Contrast of Quality Control and Two Levels—Workforce and Upper Management

	At Workforce Levels	At Managerial Levels
Control goals	Product and process features in specifications and procedures	Business-oriented, product salability, competitiveness
Sensors	Technological	Data systems
Decisions to be made	Conformance or not?	Meet customer needs or not?

Source: "Making Quality Happen," Juran Institute, Inc., Senior Executive Workshop (1988), p. F-4, Southbury, Conn.

Planning for Control

Planning for control is the activity that provides the system—the concepts, methodology, and tools—through which company personnel can keep the operating processes stable and thereby produce the product features required to meet customer needs. The input-output features of this system (also plan, process) were depicted in Figure 7.2.

Critical to Quality (CTQ): Customers and Their Needs

The principal customers of control systems are the company personnel engaged in control—those who carry out the steps that enable the feedback loop. Such personnel require (1) an understanding of what is critical to quality (CTQ) and customers' needs and (2) a definition of their own role in meeting those needs. However, most of them lack direct contact with customers. Planning for control helps to bridge that gap by supplying a translation of what customers' needs are, along with defining responsibility for meeting those needs. In this way, planning for quality control includes providing operating personnel with information on customer needs (whether direct or translated) and defining the related control responsibilities of the operating personnel. Planning for quality control can run into extensive detail.

Who plans for control? Planning for control has in the past been assigned to:

- Product development staff
- Quality engineers and specialists

▲ Multifunctional design teams
▲ Departmental leaders and supervisors
▲ The workforce

Planning for control of critical processes has traditionally been the responsibility of those who plan the operating process. For noncritical processes, the responsibility was usually assigned to quality specialists from the quality department. Their draft plans were then submitted to the operating heads for approval.

Recent trends have been to increase the use of the team concept. The team membership includes the operating forces and may also include suppliers and customers of the operating process. The recent trend has also been to increase participation by the workforce.

Compliance and Control Concepts

The methodologies of compliance and control are built around various concepts, such as the feedback loop, process capability, self-control, etc. Some of these concepts are of ancient origin; others have evolved in this century and in earlier centuries. During the discussion of planning for control, we will elaborate on some of the more widely used concepts.

Process Capability

One of the most important concepts in the quality planning process is *process capability*. The prime application of this concept occurs during planning of the operating processes.

This same concept also has applications in quality control. To explain this, a brief review is in order. All operating processes have an inherent uniformity for producing products. This uniformity can often be quantified, even during the planning stages. The process planners can use the resulting information for making decisions on adequacy of processes, choice of alternative processes, need for revision of processes, and so forth, with respect to the inherent uniformity and its relationship to process goals.

Applied to planning for quality control, the state of process capability becomes a major factor in decisions on frequency of measuring process performance, scheduling maintenance of facilities, etc. The greater the stability and uniformity of the process, the less the need for frequent measurement and maintenance.

Those who plan for quality control should have a thorough understanding of the concept of process capability and its application to both areas of planning—planning the operating processes as well as planning the controls.

Process Conformance

Does the process conform to its quality goals? The umpire answers this question by interpreting the observed differences between process performance and process goals. When current performance does differ from the quality goals, the question arises, What is the cause of this difference?

Special and Common Causes of Variation

Observed differences usually originate in one of two ways: (1) the observed change is caused by the behavior of a major variable in the process (or by the entry of a new major variable) or (2) the observed change is caused by the interplay of multiple minor variables in the process.

Shewhart called (1) and (2) *assignable* and *nonassignable* causes of variation, respectively (Shewhart, 1931). Deming later coined the terms *special* and *common* causes of variation (Deming, 1986). In what follows we will use Deming's terminology.

Special causes are typically sporadic and often have their origin in single variables. For such cases, it is comparatively easy to conduct a diagnosis and provide remedies. *Common causes* are typically chronic and usually have their origin in the interplay among multiple minor variables. As a result, it is difficult to diagnose them and to provide remedies. This contrast makes clear the importance of distinguishing special causes from common causes when interpreting differences. The need for making such distinctions is widespread. Special causes are the subject of quality control; common causes are the subject of quality improvement.

The Shewhart Control Chart

It is most desirable to provide umpires with tools that can help to distinguish between special causes and common causes. An elegant tool for this

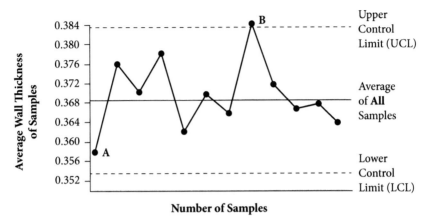

Figure 7.8 Shewhart control chart. (From "Quality Control," *Leadership for the Quality Century*, Juran Institute, Inc.)

purpose is the Shewhart control chart (or just control chart) shown in Figure 7.8.

In Figure 7.8, the horizontal scale is time and the vertical scale is quality performance. The plotted points show quality performance as time progresses.

The chart also exhibits three horizontal lines. The middle line is the average of past performance and is, therefore, the expected level of performance. The other two lines are statistical *limit lines*. They are intended to separate special causes from common causes, based on some chosen level of probability, such as 1 chance in 100.

Points Within Control Limits

Point A on the chart differs from the historical average. However, since point A is within the limit lines, this difference could be due to common causes (at a probability of more than 1 in 100). Hence, we assume that there is no special cause. In the absence of special causes, the prevailing assumptions include these:

- Only common causes are present.
- The process is in a state of *statistical control*.
- The process is doing the best it can.
- The variations must be endured.

▲ No action need be taken—taking action may make matters worse (a phenomenon known as *hunting* or *tampering*).

The preceding assumptions are being challenged by a broad movement to improve process uniformity. Some processes exhibit no points outside of control chart limits, yet the interplay of minor variables produces some defects.

In one example, a process in statistical control was nevertheless improved by an order of magnitude. The improvement was by a multifunctional improvement team, which identified and addressed some of the minor variables. This example is a challenge to the traditional assumption that variations due to common causes must be endured (Pyzdek, 1990).

In other cases the challenge is more subtle. There are again no points outside the control limits, but in addition, no defects are being produced. Nevertheless, the customers demand greater and greater uniformity. Examples are found in business processes (precision of estimating) as well as in manufacturing (batch-to-batch uniformity of chemicals, uniformity of components going into random assembly). Such customer demands are on the increase, and they force suppliers to undertake projects to improve the uniformity of even the minor variables in the process. There are many types of control charts.

Points Outside of Control Limits

Point B also differs from the historical average, but is outside of the limit lines. Now the probability is against this being the result of common causes, less than 1 chance in 100. Hence, we assume that point B is the result of special causes. Traditionally, such "out-of-control" points become nominations for corrective action.

Ideally, all such nominations should stimulate prompt corrective action to restore the status quo. In practice, many out-of-control changes do not result in corrective action. The usual reason is that the changes involving special causes are too numerous—the available personnel cannot deal with all of them. Hence, priorities are established based on economic significance or on other criteria of importance. Corrective action is taken for the high-priority cases; the rest must wait their turn. Some changes at low levels of priority may wait a long time for corrective action.

A further reason for failure to take corrective action is a lingering confusion between statistical control limits and quality tolerances. It is easy to be carried away by the elegance and sensitivity of the control chart. This happened on a large scale during the 1940s and 1950s. Here are two examples from my personal experience:

- A large automotive components factory placed a control chart at every machine.
- A viscose yarn factory created a "war room" of more than 400 control charts.

In virtually all such cases the charts were maintained by the quality departments but ignored by the operating personnel. Experience with such excesses has led leaders and planners to be wary of employing control charts just because they are sensitive detectors of change. Instead, the charts should be justified based on value added. Such justifications include these:

- Customer needs are directly involved.
- There is risk to human safety or the environment.
- Substantial economics are at stake.
- The added precision is needed for control.

Statistical Control Limits and Tolerances

For most of human history, targets and goals consisted of product features or process features, usually defined in words. Phrases such as "the color is red" and "the length is long enough" are targets, but are open to too much interpretation. The growth of technology stimulated the growth of measurement, plus a trend to define targets and goals in precise numbers. In addition, there emerged the concept of limits, or "tolerances," around the targets and goals. For example:

- Ninety-five percent of the shipments shall meet the scheduled delivery date.
- The length of the bar shall be within 1 mm of the specified number.
- The length of time to respond to customers is 10 minutes, plus or minus 2 minutes.

Such targets had official status. They were set by product or process designers and published as official specifications. The designers were the official quality legislators—they enacted the laws. Operating personnel were responsible for obeying the quality laws—meeting the specified goals and tolerances.

Statistical control limits in the form of control charts were virtually unknown until the 1940s. At that time, these charts lacked official status. They were prepared and published by quality specialists from the quality department. To the operating forces, control charts were a mysterious, alien concept. In addition, the charts threatened to create added work in the form of unnecessary corrective action. The operating personnel reasoned as follows: It has always been our responsibility to take corrective action whenever the product becomes nonconforming. These charts are so sensitive that they detect process changes that do not result in nonconforming products. We are then asked to take corrective action even when the products meet the quality goals and tolerances.

So there emerged a confusion of responsibility. The quality specialists were convinced that the control charts provided useful early-warning signals that should not be ignored. Yet the quality departments failed to recognize that the operating forces were now faced with a confusion of responsibility. The latter felt that there was no need for corrective action as long as the products met the quality goals. The upper leaders of those days were of no help—they did not involve themselves in such matters. Since the control charts lacked official status, the operating forces solved their problem by ignoring the charts. This contributed to the collapse in the 1950s of the movement known as *statistical quality control.*

The 1980s created a new wave of interest in applying the tools of statistics to the control of quality. Many operating personnel underwent training in *statistical process control.* This training helped to reduce the confusion, but some confusion remains. To get rid of the confusion, leaders should:

▲ Clarify the responsibility for corrective action on points outside the control limits. Is this action mandated or is it discretionary?
▲ Establish guidelines on actions to be taken when points are outside the statistical control limits but the product still meets the quality tolerances.

The need for guidelines for decision making is evident from Figure 7.9. The guidelines for quadrants A and C are obvious. If both process and

		Product	
		Conforms	Does Not Conform
Process	Does Not Conform	B Vague	C Clear
	Conforms	A Clear	D Vague

Figure 7.9 Areas of decision making. (From "Making Quality Happen," Juran Institute, Inc., 1988. Used by permission.)

product conform to their respective goals, the process may continue to run. If neither process nor product conforms to its respective goals, the process should be stopped and remedial action should be taken. The guidelines for quadrants B and D are often vague, and this vagueness has been the source of a good deal of confusion. If the choice of action is delegated to the workforce, the leaders should establish clear guidelines.

Numerous efforts have been made to design control chart limits in ways that help operating personnel detect whether product quality is threatening to exceed the product quality limits.

Self-Control and Controllability

Workers are in a state of self-control when they have been provided with all the essentials for doing good work. These essentials include

- Means of knowing what the goals are.
- Means of knowing what their actual performance is.
- Means for changing their performance in the event that performance does not conform to goals. To meet this criterion requires an operating process that (1) is inherently capable of meeting the goals and (2) is

provided with features that make it possible for the operating forces to adjust the process as needed to bring it into conformance with the goals.

These criteria for self-control are applicable to processes in all functions and at all levels, from general manager to nonsupervisory worker.

It is all too easy for leaders to conclude that the above criteria have been met. In practice, however, there are many details to be worked out before the criteria can be met. The nature of these details is evident from checklists, which have been prepared for specific processes to ensure that the criteria for self-control are met. Examples of these checklists include those designed for product designers, production workers, and administrative and support personnel.

If all the criteria for self-control have been met at the worker level, any resulting product nonconformances are said to be *worker-controllable*. If any of the criteria for self-control have not been met, then management's planning has been incomplete—the planning has not fully provided the means for carrying out the activities within the feedback loop. The nonconforming products resulting from such deficient planning are then said to be *management-controllable*. In such cases it is risky for leaders to hold the workers responsible for quality.

Responsibility for results should, of course, be keyed to controllability. However, in the past, many leaders were not aware of the extent of controllability as it prevailed at the worker level. Studies conducted by Juran during the 1930s and 1940s showed that at the worker level, the proportion of management-controllable to worker-controllable nonconformances was of the order of 80:20. These findings were confirmed by other studies during the 1950s and 1960s. That ratio of 80:20 helps to explain the failure of so many efforts to solve the organizations' quality problems solely by motivating the workforce.

Effect on the Process Conformance Decision

Ideally, the decision of whether the process conforms to process quality goals should be made by the workforce. There is no shorter feedback loop. For many processes, this is the actual arrangement. In other cases, the process conformance decision is assigned to nonoperating personnel—independent checkers or inspectors. The reasons include these:

- The worker is not in a state of self-control.
- The process is critical to human safety or to the environment.
- Quality does not have top priority.
- There is a lack of mutual trust between the leaders and the workforce.

Product Conformance: Fitness for Purpose

There are two levels of product features, and they serve different purposes. One of these levels serves such purposes as:

- Meeting customer needs
- Protecting human safety
- Protecting the environment

Product features are said to possess *fitness for use* if they are able to serve the above purposes.

The second level of product features serves purposes such as:

- Providing working criteria to those who lack knowledge of fitness for use
- Creating an atmosphere of law and order
- Protecting innocents from unwarranted blame

Such product features are typically contained in internal specifications, procedures, standards, etc. Product features that are able to serve the second list of purposes are said to possess conformance to specifications, etc. We will use the shorter label *conformance*.

The presence of two levels of product features results in two levels of decision making: Is the product in conformance? Is the product fit for use? Figure 7.10 shows the interrelation of these decisions to the flow diagram.

The Product Conformance Decision

Under prevailing policies, products that conform to specification are sent on to the next destination or customer. The assumption is that products

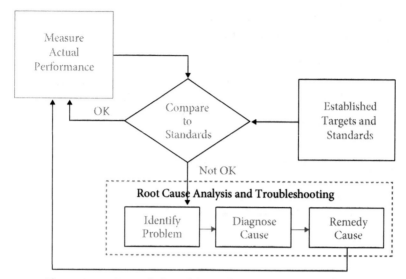

Figure 7.10 Interrelation of QC and root cause analysis (RCA).

that conform to specification are also fit for use. This assumption is valid in the great majority of cases.

The combination of large numbers of product features, when multiplied by large volumes of product, creates huge numbers of product conformance decisions to be made. Ideally, these decisions should be delegated to the lowest levels of organization—to the automated devices and the operating workforce. Delegation of this decision to the workforce creates what is called *self-inspection*.

Self-Inspection

We define *self-inspection* as a state in which decisions on the product are delegated to the workforce. The delegated decisions consist mainly of these questions: Does product quality conform to the quality goals? What disposition is to be made of the product?

Note that self-inspection is very different from self-control, which involves decisions on the *process*.

The merits of self-inspection are considerable:

- The feedback loop is short; the feedback often goes directly to the actuator—the energizer for corrective action.

- Self-inspection enlarges the job of the workforce—it confers a greater sense of job ownership. Self-inspection removes the police atmosphere created by use of inspectors, checkers, etc.

However, to make use of self-inspection requires meeting several essential criteria:

- *Quality is number one.* Quality must undoubtedly be the top priority.
- *Mutual confidence.* The leaders must have enough trust in the workforce to be willing to make the delegation, and the workforce must have enough confidence in the leaders to be willing to accept the responsibility.
- *Self-control.* The conditions for self-control should be in place so that the workforce has all the means necessary to do good work.
- *Training.* The workers should be trained to make the product conformance decisions.
- *Certification.* The recent trend is to include a certification procedure. Workers who are candidates for self-inspection undergo examinations to ensure that they are qualified to make good decisions. The successful candidates are certified and may be subject to audit of decisions thereafter.

In many organizations, these criteria are not fully met, especially the criterion of priority. If some parameter other than quality has top priority, there is a real risk that evaluation of product conformance will be biased. This problem arises frequently when personal performance goals are in conflict with overall quality goals. For example, a chemical company found that it was rewarding sales personnel on revenue targets without regard to product availability or even profitability. The salespeople were making all their goals, but the company was struggling.

The Fitness for Purpose Decision

The great majority of products do conform to specifications. For the nonconforming products there arises a new question: Is the nonconforming product nevertheless fit for use?

A complete basis for making this decision requires answers to questions such as the following:

- Who are the users?
- How will this product be used?
- Are there risks to structural integrity, human safety, or the environment?
- What is the urgency for delivery?
- How do the alternatives affect the producer's and the user's economics?

To answer such questions can require considerable effort. Organizations have tried to minimize the effort through procedural guidelines. The methods in use include these:

- *Treat all nonconforming products as unfit for use.* This approach is widely used for products that can pose risks to human safety or the environment—products such as pharmaceuticals or nuclear energy.
- *Create a mechanism for decision making.* An example is the material review board so widely used in the defense industry. This device is practical for matters of importance, but is rather elaborate for the more numerous cases in which little is at stake.
- *Create a system of multiple delegation.* Under such a system, the vital few decisions are reserved for a formal decision-making body such as a material review board. The rest are delegated to other people.

Table 7.3 is an example of a table of delegation used by a specific company (personal communication to one of the authors.)

Disposition of Unfit Product

Unfit product is disposed of in various ways: scrap, sort, rework, return to supplier, sell at a discount, etc. The internal costs can be estimated to arrive at an economic optimum. However, the effects go beyond money: schedules are disrupted, people are blamed, etc. To minimize the resulting human abrasion, some organizations have established rules of conduct, such as these:

- *Choose the alternative that minimizes the total loss to all parties involved.* Now there is less to argue about, and it becomes easier to agree on how to share the loss.
- *Avoid looking for blame.* Instead, treat the loss as an opportunity for quality improvement.

Table 7.3 Multiple Delegations of Decision Making on Fitness for Purpose*

Effect of Nonconformance Is On	Amount of Product or Money at Stake Is:	
	Small	Large
Internal economics only	Department head directly involved, quality engineer	Plant managers involved, quality manager
Economic relations with supplier	Supplier, purchasing agent, quality engineer	Supplier, manager
Economic relations with client	Client, salesperson, quality engineer	Client (for marketing, manufacturing, technical, quality)
Field performance of the product	Product designer, salesperson, quality engineer	Client (managers for technical, manufacturing, marketing, quality)
Risk of damage to society or of nonconformance to government regulations	Product design manager, compliance officer, lawyer, quality managers	General manager and team of upper managers

*For those industries whose quality mission is really one of conformance to specification (for example, atomic energy, space), the real decision-maker of fitness for use is the client or the government regulator.

▲ *Use chargebacks sparingly.* Charging the vital few losses to the departments responsible has merit from an accounting viewpoint. However, when applied to the numerous minor losses, this is often uneconomic as well as detrimental to efforts to improve quality.

Failure to use products that meet customer needs is a waste. Sending out products that do not meet customer needs is worse. Personnel who are assigned to make product conformance decisions should be provided with clear definitions of responsibility as well as guidelines for decision making. Leaders should, as part of their audit, ensure that the processes for making product conformance decisions are appropriate to company needs.

Corrective Action

The final step in closing the feedback loop is to actuate a change that restores conformance with quality goals. This step is popularly known as *troubleshooting* or *firefighting*.

Note that the term *corrective action* has been applied loosely to two very different situations, as shown in Figure 7.1. The feedback loop is well designed to eliminate sporadic nonconformance, like that "spike" in Figure 7.1; the feedback loop is not well designed to deal with the area of chronic waste shown in the figure. Instead, the need is to employ the quality improvement process. We will use the term *corrective action* in the sense of troubleshooting—eliminating sporadic nonconformance.

Corrective action requires the journeys of diagnosis and remedy. These journeys are similar to quality improvement. Sporadic problems are the result of adverse change, so the diagnostic journey aims to discover what has changed. The remedial journey aims to remove the adverse change and restore conformance.

Diagnosing Sporadic Change

During the diagnostic journey, the focus is on what has changed. Sometimes the causes are not obvious, so the main obstacle to corrective action is diagnosis. The diagnosis makes use of methods and tools such as the following:

- Forensic autopsies to determine with precision the symptoms exhibited by the product and process.
- Comparison of products made before and after the trouble began to see what has changed; also comparison of good and bad products made since the trouble began.
- Comparison of process data before and after the problem began to see what process conditions have changed.
- Reconstruction of the chronology, which consists of logging on a time scale (of hours, days, etc.) (1) the events that took place in the process before and after the sporadic change—that is, rotation of shifts, new employees on the job, maintenance actions, etc., and (2) the time related to product information—that is, date codes, cycle time for processing, waiting time, move dates, etc.

Analysis of the resulting data usually sheds a good deal of light on the validity of the various theories of causes. Certain theories are denied. Other theories survive to be tested further.

Operating personnel who lack the training needed to conduct such diagnoses may be forced to shut down the process and request assistance from specialists, the maintenance department, etc. They may also run the process "as is" in order to meet schedules and thereby risk failure to meet the quality goals.

Corrective Action—Remedy

Once the cause(s) of the sporadic change is (are) known, the worst is over. Most remedies consist of going back to what was done before. This is a return to the familiar, not a journey into the unknown (as is the case with chronic problems). The local personnel are usually able to take the necessary action to restore the status quo.

Process designs should provide the means to adjust the process as required to attain conformance with quality goals. Such adjustments are needed at start-up and during the running of the process. This aspect of design for process control ideally should meet the following criteria:

- There should be a known relationship between the process variables and the product results.
- Means should be provided for ready adjustment of the process settings for the key process variables.
- A predictable relationship should exist between the amount of change in the process settings and the amount of effect on the product features.

If such criteria are not met, the operating personnel will, in due course, be forced to cut corners to carry out remedial action. The resulting frustrations become a disincentive to putting a high priority on quality.

The Role of Statistical Methods in Control

An essential activity within the feedback loop is the collection and analysis of data. This activity falls within the scientific discipline known as *statistics*. The methods and tools used are often called *statistical methods*. These methods have long been used to aid in data collection and analysis in many fields: biology, government, economics, finance, management, etc.

Statistical Process Control (SPC)

The term has multiple meanings, but in most organizations, statistical process control is considered to include basic data collection; analysis through such tools as frequency distributions, Pareto principle, Ishikawa (fish bone) diagram, Shewhart control chart, etc.; and application of the concept of process capability.

Advanced tools, such as design of experiments and analysis of variance, are a part of statistical methods but are not normally considered part of statistical process control.

The Merits

These statistical methods and tools have contributed in an important way to quality control and to the other processes of the Juran Trilogy—quality improvement and quality planning. For some types of quality problems, the statistical tools are more than useful—the problems cannot be solved at all without using the appropriate statistical tools.

The SPC movement has succeeded in training a great many supervisors and workers in basic statistical tools. The resulting increase in statistical literacy has made it possible for them to improve their grasp of the behavior of processes and products. In addition, many have learned that decisions based on data collection and analysis yield superior results.

The Risks

There is a danger in taking a tool-oriented approach to quality instead of a problem-oriented or results-oriented approach. During the 1950s, this preoccupation became so extensive that the entire statistical quality control movement collapsed; the word *statistical* had to be eliminated from the names of the departments.

The proper sequence in managing is first to establish goals and then to plan how to meet those goals, including choosing the appropriate tools. Similarly, when dealing with problems—threats or opportunities—experienced leaders start by first identifying the problems. They then try to solve those problems by various means, including choosing the proper tools.

During the 1980s, numerous organizations did, in fact, try a tool-oriented approach by training large numbers of their personnel in the use of statistical tools. However, there was no significant effect on the bottom line. The reason was that no infrastructure had been created to identify which projects to tackle, to assign clear responsibility for tackling those projects, to provide needed resources, to review progress, etc.

Leaders should ensure that training in statistical tools does not become an end in itself. One form of such assurance is through measures of progress. These measures should be designed to evaluate the effect on operations, such as improvement in customer satisfaction or product performance, reduction in cost of poor quality, etc. Measures such as numbers of courses held or numbers of people trained do not evaluate the effect on operations and hence should be regarded as subsidiary in nature.

Information for Decision Making

Quality control requires extensive decision making. These decisions cover a wide variety of subject matter and take place at all levels of the hierarchy. The planning for quality control should provide an information network that can serve all decision-makers. At some levels of the hierarchy, a major need is for real-time information to permit prompt detection and correction of nonconformance to goals. At other levels, the emphasis is on summaries that enable leaders to exercise control over the vital few control subjects. In addition, the network should provide information as needed to detect major trends, identify threats and opportunities, and evaluate the performance of organization units and personnel.

In some organizations, the quality information system is designed to go beyond control of product features and process features; the system is also used to control the quality performance of organizations and individuals, such as departments and department heads. For example, many organizations prepare and regularly publish scoreboards showing summarized quality performance data for various market areas, product lines, operating functions, etc. These performance data are often used as indicators of the quality performance of the personnel in charge.

To provide information that can serve all those purposes requires planning that is directed specifically at the information system. Such planning is best done by a multifunctional team whose mission is focused on the quality information system. That team properly includes the customers as well as the suppliers of information. The management audit of the quality control system should include assurance that the quality information system meets the needs of the various customers.

The Quality Control System and Policy Manual

A great deal of quality planning is done through *procedures*, which are really repetitive-use plans. Such procedures are thought out, written out, and approved formally. Once published, they become the authorized ways of conducting the company's affairs. It is quite common for the procedures related to managing for quality to be published collectively in a Quality Manual (or similar title). A significant part of the manual relates to quality control.

Quality manuals add to the usefulness of procedures in several ways:

- *Legitimacy*. The manuals are approved at the highest levels of the organization.
- *Easy to find*. The procedures are assembled into a well-known reference source rather than being scattered among many memoranda, oral agreements, reports, minutes, etc.
- *Stable*. The procedures survive despite lapses in memory and employee turnover.

Study of company quality manuals shows that most contain a core content, which is quite similar from company to company. Relative to quality control, this core content includes procedures for:

- Applying the feedback loop to process and product control
- Ensuring that operating processes are capable of meeting the quality goals
- Maintaining facilities and calibration of measuring instruments
- Relating to suppliers on quality matters
- Collecting and analyzing the data required for the quality information system
- Training the personnel to carry out the provisions of the manual
- Auditing to ensure adherence to procedures

The need for repetitive-use quality control systems has led to an evolution of standards at industry, national, and international levels. For elaboration, see *Juran's Quality Handbook*, 6th ed., Chapter 16, "Using International Standards to Ensure Organization Compliance."

Provision for Audits

Experience has shown that control systems are subject to "slippage" of all sorts. Personnel turnover may result in loss of essential knowledge. Entry of unanticipated changes may result in obsolescence. Shortcuts and misuse may gradually undermine the system until it is no longer effective.

The major tool for guarding against deterioration of a control system has been the audit. Under the audit concept, a periodic, independent review is established to provide answers to the following questions: Is the control system still adequate for the job? Is the system being followed?

The answers are obviously useful to the operating leaders. However, that is not the only purpose of the audit. A further purpose is to provide those answers to people who, though not directly involved in operations, nevertheless have a need to know. If quality is to have top priority, those who have a need to know include the upper leaders.

It follows that one of the responsibilities of leaders is to mandate establishment of a periodic audit of the quality control system.

Tasks for Leaders

1. Leaders should avoid getting too deeply involved in making decisions on quality control. They should make the vital few decisions, provide criteria to distinguish the vital few from the rest, and delegate the rest under a decision-making process.
2. To eliminate the confusion relative to control limits and product quality tolerance, leaders should clarify the responsibility for corrective action on points outside the control limits and establish guidelines on action to be taken when points are outside the statistical control limits but the product still meets the quality tolerances.
3. Leaders should, as part of their audit, ensure that the processes for making product conformance decisions are appropriate to company needs. They should also ensure that training in statistical tools does not become an end in itself. The management audit of the quality control system should include assurance that the quality information system meets the needs of the various customers.

4. Leaders are able to influence the adequacy of the quality control manual in several ways: participate in defining the criteria to be met, approve the final draft to make it official, and periodically audit the up-to-date aspect of the manual as well as the state of conformance.

References

Deming, W. E. (1950). *Elementary Principles of the Statistical Control of Quality*. Nippon Kagaku Gijutsu Renmei (Japanese Union of Scientists and Engineers), Tokyo.

Deming, W. E. (1986). *Out of the Crisis*. MIT Center for Advanced Engineering Study, Cambridge, Mass.

Juran, J. M. (1964). *Managerial Breakthrough*. McGraw-Hill, New York.

Juran J. M. and DeFeo J. A. (2010). *Juran's Quality Handbook: The Complete Guide to Performance Excellence*, 6th ed. McGraw-Hill, New York.

Pyzdek, T. (1990). "There's No Such Thing as a Common Cause." *ASQC Quality Congress Transactions*, pp. 102–108.

Radford, G. S. (1917). "The Control of Quality." *Industrial Management*, vol. 54, p. 100.

Radford, G. S. (1922). *The Control of Quality in Manufacturing*. Ronald Press Company, New York.

Radman, M., and R. Wagner. (1988). "The High Fidelity of DNA Duplication." *Scientific American*, August, pp. 40–46.

Shewhart, W. A. (1931). *Economic Control of Quality of Manufactured Product*. Van Nostrand, New York. Reprinted by ASQC, Milwaukee, 1980.

CHAPTER 8

Simplifying Macro Processes with Business Process Management

Success in achieving superior results depends heavily on managing such large, complex, multifunctional business processes as product development, the revenue cycle, invoicing, patient care, purchasing, materials procurement, supply chain, and distribution, among others. In the absence of management's attention over time, many processes may become too slow, obsolete, overextended, redundant, excessively costly, ill defined, and not adaptable to the demands of a constantly changing environment. For processes that have suffered this neglect, quality of the output falls far short of the quality required for competitive performance. This chapter focuses on helping an organization simplify and sustain its performance through process ownership of important business processes. Business process ownership happens after an organization masters all processes of the Juran Trilogy.

Why Business Process Management?

The dynamic environment in which business is conducted today is characterized by what has been referred to as the *six C's: change, complexity, customer demands, competitive pressure, cost impacts,* and *constraints*. All have a great impact on an organization's ability to meet its stated business goals and objectives. Organizations have responded to these factors by developing new products and services. They also have carried out numerous breakthrough projects and are at a maturity level conducive to process ownership.

A business process is the logical organization of people, materials, energy, equipment, and information into work activities designed to produce a required end result (product or service).

PALL (1987)

There are three principal dimensions for measuring process performance: effectiveness, efficiency, and adaptability:

1. The process is *effective* if the output meets customer needs.
2. It is *efficient* when it is effective at the least cost.
3. The process is *adaptable* when it remains effective and efficient in the face of the many changes that occur over time.

On the surface, the need to maintain high-quality processes would seem obvious. To understand why good process quality is the exception and not the rule requires us to look closely at how processes are designed and what happens to them over time.

Business process management (BPM) has become a critical component of information technology (IT) programs. Without having good business process management system, an IT system can fail. All disciplined IT implementations must include well-developed BPM processes. With technology, BPM allows organizations to abstract business processes from the technology infrastructure and go far beyond automating business processes or solving business problems. BPM enables business to respond to changing consumer, market, and regulatory demands faster than competitors, creating a competitive advantage. In the IT world, BPM is often called a *BPM life cycle*.

For reasons of history, the business organization model has evolved into a hierarchy of functionally specialized departments. Management direction, goals, and measurements are deployed from the top downward through this vertical hierarchy. However, the processes that yield the products of work—in particular, products that customers buy (and that justify the existence of the organization)—flow horizontally across the organization through functional departments (Figure 8.1). Traditionally, each functional piece of a process is the responsibility of a department, whose manager is held accountable for the performance of that piece. However, no one is accountable for the entire process. Many problems arise from the

Chapter 8 Simplifying Macro Processes with Business Process Management | 213

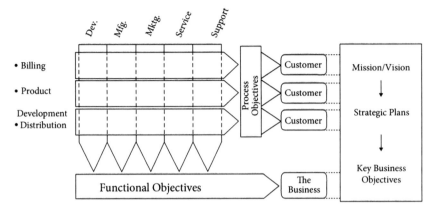

Figure 8.1 Horizontal flow through functional departments.

conflict between the demands of the departments and the demands of the overall major processes.

In a competition with functional goals, functional resources, and functional careers, cross-functional processes are starved for attention. As a result, the processes as operated are often neither effective nor efficient, and they are certainly not adaptable.

A second source of poor process performance is the natural deterioration to which all processes are subject in the course of their evolution. For example, at one railroad, the company telephone directory revealed that there were more employees with the title Rework Clerk than with the title Clerk. Each of the Rework Clerks had been put in place to guard against the recurrence of some serious problem that arose. Over time, the imbalance in titles was the outward evidence of processes that had established rework as the organization's norm.

The rapidity of technological evolution, in combination with rising customer expectations, has created global competitive pressures on costs and quality. These pressures have stimulated an exploration of cross-functional processes—to identify and understand them and to improve their performance. There is now much evidence that, within the total product cycle, a major problem of poor process performance lies with BPM technologies. Functional objectives frequently conflict with customer needs, served as they must be by cross-functional processes. Furthermore, the processes generate a variety of waste (e.g., missed deadlines, factory scrap). It is not difficult to identify such products—generating invoices, preparing insurance policies, or paying a claim—that take over 20 days to accomplish in less than 20 minutes of actual work. Processes are also not easily changed in response to the continuously changing environment. To better serve customer needs, there is a need to restore these processes to effectiveness, efficiency, and adaptability.

The BPM Methodology

BPM is initiated when executive management selects key processes, identifies owners and teams, and provides them with process goal statements. After the owners and the team are trained in performance excellence methods and tools, they work through the three phases of BPM methodology: planning, transfer, and operational management.

The *planning phase,* in which the process design (or redesign) takes place and is the most time-consuming of the three phases, involves five steps:

1. Define the present process.
2. Determine customer needs and process flow.
3. Establish process measurements.
4. Conduct analyses of measurement and other data.
5. Design the new process. The output is the new process plan.

The *transfer phase* is the second phase, in which the plans developed in the first phase are handed off from the process team to the operating forces and put into operation.

The *operational management phase* is the third phase of BPM. Here, the working owner and team first monitor new process performance, focusing on process effectiveness and efficiency measurements. They apply quality control techniques, as appropriate, to maintain performance. They use quality improvement techniques to rid the process of chronic deficiencies. Finally, they conduct a periodic executive management review and assessment to ensure that the process continues to meet customer and business needs and remains competitive.

Note: BPM is not a one-time event; it is itself a continuous process carried out in real time.

Deploying BPM

Selecting Key Macro Process(es)

Organizations operate dozens of major cross-functional business processes. From these, a few key processes are selected as the BPM focus. An organization's strategic plan provides guidance in selecting key processes. There are several approaches to doing so:

- The *critical success factor* approach holds that, for any organization, relatively few (no more than eight) factors can be identified as "necessary and sufficient" for attaining its mission and vision. Once identified, these factors are used to select the key business processes and rank them by priority (Hardaker and Ward, 1987).
- The *balanced business scorecard* (Kaplan and Norton, 1992) measures business performance in four dimensions: financial performance, performance in the eyes of the customer, internal process performance, and performance in organization learning and innovation. Performance measures are created and performance targets are set for each dimension. Using these measures to track performance provides a *balanced* assessment of business performance. Processes that create imbalances in the scorecard are identified as processes that need attention most—the key processes.
- Another approach is to invite upper management to identify a few (four to six) organization-specific critical selection criteria to use in evaluating the processes. Examples of such criteria are the effect on business success, the effect on customer satisfaction, the significance of problems associated with the process, the amount of resources currently committed to the process, the potential for improvement, the affordability of adopting BPM, and the effect of the process on the schedule. Using these criteria and a simple scoring system (such as low, medium, or high), managers evaluate the many processes from the long list of the organization's major business processes (10 to 25 of them) and, by comparing the evaluations, identify the key processes. (The long list may be prepared in advance in a process identification study conducted separately, often by the chief quality officer and with the support of a consultant.)

Whatever approach is used to identify key processes, the process map can be used to display the results. The *process map* is a graphical tool for describing an organization in terms of its business processes and their relationships to the organization's principal stakeholders. The traditional organization chart answers the question, Who reports to whom? The process map answers the question, How does the organization's work get done?

Organizing for BPM

Because certain major cross-functional business processes, the *key processes*, are critical to business success, the quality council sees to it that those processes are organized in a special way. After selecting key processes, the quality council appoints a process owner, who is responsible for making the process effective, efficient, and adaptable and is accountable for its performance (Riley, 1989; Riley et al., 1994).

For large complex processes, especially in large organizations, a two-tier ownership arrangement is used most often. An appointed executive owner operates as a sponsor, champion, and supporter at the upper management level and is accountable for process results. At the operating level, a working owner, usually a first- or second-level manager, leads the process management team responsible for day-to-day operation. Owner assignments—executive owner and working owner—are ongoing. The major advantages of this structure are that there is, at the same time, "hands-on" involvement and support of upper management and adequate management of the process details.

The process management team is a peer-level group that includes a manager or supervisor from each major function within the process. Each member is an expert in a segment of the process. Ideally, BPM teams have no more than eight members, and the individuals chosen should be proven leaders. The team is responsible for managing and continuously improving the process. The team shares with the owner the responsibilities for effectiveness and efficiency. Most commonly, team assignments are ongoing.

From time to time, a process owner creates an ad hoc team to address some special issue (human resources, information technology, activity-based costing, etc.). The mission of such a project-oriented team is limited, and the team disbands when the mission is complete. The ad hoc team is different from the process management team.

Chapter 8 Simplifying Macro Processes with Business Process Management | 217

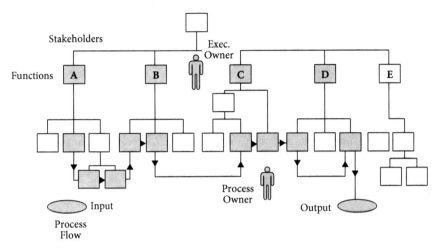

Figure 8.2 Diagram of a multifunctional organization and one of its major processes.

Figure 8.2 is a simplified diagram of a multifunctional organization and one of its major processes. The shaded portions include the executive owner, the working owner, the BPM team, and the stakeholders—functional heads at the executive level who have work activities of the business process operating within their function. Customarily, stakeholders are members of the quality council, along with the executive owner. Taken together, this shaded portion is referred to as the BPM Infrastructure.

Establishing the Team's Mission and Goals

The preliminary process mission and improvement goals for the process are communicated to the owners (executive and working levels) and the team by the quality council. To do their jobs most effectively, the owners and the team must make the mission and goals their own. They do this by defining the process, the first step of the planning phase.

The Planning Phase: Planning the New Process

The first phase of BPM is planning, which consists of five steps: (1) defining the process, (2) discovering customer needs and flowcharting the process, (3) establishing measurements of the process, (4) analyzing process measurements and other data, and (5) designing (or redesigning) the process. The output of the planning phase is the new process plan.

Defining the Current Process

The owner(s) and the team collaborate to define the process precisely. In accomplishing this, the starting point and principal reference is the process documentation developed by the quality council during the selection of key processes and identification of owners and teams. This documentation includes preliminary statements of mission and goals.

Effective mission and goal statements explicitly declare:

- The purpose and scope of the process.
- "Stretch" targets for customer needs and business needs.

The purpose of the stretch target is to motivate aggressive process improvement activity. For example, a mission statement for the special-contract management process is to provide competitive special pricing and supportive terms and conditions for large information system procurements that meet customer needs for value, contractual support, and timeliness at affordable cost.

The goals for the same process are to:

- Deliver approved price and contract support document within 30 days of date of customer's letter of intent.
- Achieve a yield of special-contract proposals (percentage of proposals closed as sales) of not less than 50 percent.

The team must reach a consensus on the suitability of these statements, propose modifications for the quality council's approval, if necessary, and document the scope, objectives, and content. Based on available data and collective team experience, the team will document process flow, process strengths and weaknesses, performance history, measures, costs, complaints, environment, and resources. This will probably involve narrative documentation and will certainly require the use of flow diagrams.

Bounding the business process starts with inventorying the major subprocesses—six to eight of them is typical—that the business process comprises. The inventory must include the "starts with" subprocess (the first subprocess executed), the "ends with" subprocess (the last executed), and the major subprocesses in between. If they have a significant effect on the quality of the process output, activities upstream of the process are included within the process boundary. To provide focus and avoid ambiguity, it is also helpful to list subprocesses that are explicitly excluded from the business process. The accumulating information on process components is represented in diagram form, which evolves from a collection of subprocesses to a flow diagram as the steps of the planning phase are completed.

Figure 8.3 shows a high-level diagram of the special-contract (SC) process that resulted from process analysis but before the process was redesigned. At the end of the process definition step, such a diagram is not yet a flow diagram, as there is no indication of the sequence in which the subprocesses occur. Establishing those relationships as they presently exist is the work of step 2.

Discovering Customer Needs and Mapping the Current State

For the process to work well, the team must identify all customers, determine their needs, and prioritize the information. Priorities enable the team to focus its attention and spend its energies where they will be most effective.

Determining customer needs and expectations requires ongoing, disciplined activity. Process owners must ensure that this activity is incorporated in the day-to-day conduct of the business process as the customer requirements subprocess and must assign accountability for its performance. The output of this vital activity is a continually updated customer requirement statement.

On the process flowchart, it is usual to indicate key suppliers and customers and their roles in the process as providers or receivers of materials, product, information, and the like. Although the diagram can serve a number of specialized purposes, the most important here is to create a common, high-level understanding among the owner and the team members of how the process works—how the subprocesses relate to one another and to the customers and suppliers and how information and product move around and through the process. In creating the process flowchart, the team will also verify the list of customers and may, as understanding of the process deepens, add to the list of customers.

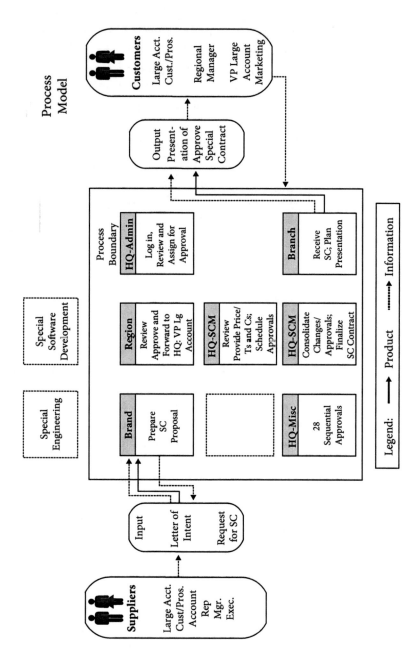

Figure 8.3 High-level diagram of the special-contract process.

The process flowchart is the team's primary tool for analyzing the process to determine whether it can satisfy customer needs. By walking through the chart together, step by step, sharing questions and collective experience, the team determines whether the process is correctly represented, making adjustments to the diagram as necessary to reflect the process as it presently operates.

When this step is complete, the team has a starting point for analyzing and improving the process. In Figure 8.4, the product flow is shown by solid lines and the information flow by dotted lines.

Establishing Process Measurements

What gets measured gets done. Establishing, collecting, and using the correct measures are critical in managing business process quality. *Process capability, process performance,* and other process measures have no practical significance if the process they purport to describe is not managed. To be managed, the process must fulfill certain minimum conditions:

1. It has an owner.
2. It is defined.
3. Its management infrastructure is in place.
4. Its requirements are established.
5. Its measurements and control points are established.
6. It demonstrates stable, predictable, and repeatable performance.

A process that fulfills these minimum conditions is said to be *manageable.* Manageability is the precondition for all further work in BPM.

Of these criteria, items 1 through 4 have already been addressed in this chapter. Criteria items 5 and 6 are addressed as follows.

Process Measurements

In deciding what aspects of the process to measure, we look for guidance to the process mission and to our list of customer needs. Process measures based on customer needs provide a way of measuring process effectiveness. For example, if the customer requires delivery of an order within 24 hours of order placement, we incorporate into our order fulfillment process a measure such as "time elapsed between receipt of order and delivery of

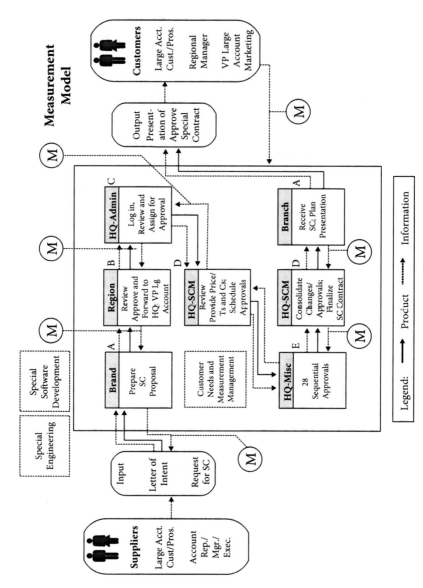

Figure 8.4 Flowchart of the special-contract process including process control points.

order" and a system for collecting, processing, summarizing, and reporting information from the data generated. The statistic reported to the executive owner will be one such as "percentage of orders delivered within 24 hours," a statistic that summarizes on-time performance. The team will also need data on which to base analysis and correction of problems and continuous improvement of the process. For this purpose, the team needs data from which they can compute such descriptive statistics as distribution of delivery times by product type, among others. The uses to which the data will be put must be thought through carefully at the time of process design to minimize redesign of the measures and measurement systems.

Process measures based on cost, cycle time, labor productivity, and process yield measure process efficiency. Suppose that a goal for our order fulfillment process is to reduce order-picking errors to 1 error per 1000 order lines. Managing that goal requires identifying order-picking errors in relation to the number of order lines picked. For inadvertent order-picking errors—that is, when they happen, the picker is unaware of them—measuring them requires a separate inspection to identify errors. In a random audit on a sample of picked orders, an inspector identifies errors and records them. As with delivery-time measurement, the team must think through all the uses it will make of these measurements. To report an estimated error rate, the data needed are the number of errors and the number of order lines inspected. To improve process performance in this category, the data must help the team identify error sources and determine their root cause. For that to occur, each error must be associated with time of day, shift, product type, and size of package so that the data can be stratified to test various theories of root cause.

Although process adaptability is not a measurement category, it is an important consideration for process owners and teams. Adaptability is discussed later in the chapter.

Process measurements must be linked to business performance. If certain key processes must run exceptionally well to ensure organization success, it follows that collective success of the key processes is good for the organization's performance. Process owners must take care to select process measures that are strongly correlated with traditional business indicators, such as revenue, profit, return on investment, earnings per share, productivity per employee, and so on. In high-level business plan reviews, managers are motivated and rewarded for maintaining this linkage between process and organization performance measures because of the two values that BPM supports: organization success is good, and BPM is the way we will achieve organization success.

Table 8.1 shows some typical process measurements and the traditional business indicators with which they are linked. To illustrate, "percent of sales quota achieved" is a traditional business indicator relating to the business objective of improving revenue. The special-contract management process has a major impact on the indicator, as more than 30 percent of U.S. revenue comes from that process. Therefore, the contract close rate (ratio of the value of firm contracts to the total value of proposals submitted) of the special-contract management process is linked to percent of sales quota and other traditional revenue measures, and is, therefore, a measure of great importance to management. Measurement points appear on the process flow diagram.

Table 8.1 Typical Process Measurements and the Traditional Business Indicators

The Traditional Business View		The Process View	
Business Objective	**Business Indicator**	**Key Process**	**Process Measure**
Higher revenue	Percent of sales quota achieved	Contract management	Contract close rate
	Percent of revenue plan achieved	Product development	Development cycle time
	Value of orders cancelled after shipment	Account management	Backlog management and system assurance timeliness
	Receivable days outstanding		Billing quality index
Reduce costs	S, G & A Inventory turns	Manufacturing	Manufacturing cycle time

Control Points

Process measurement is also a part of the control mechanisms established to maintain planned performance in the new process. To control the process requires that each of a few selected process variables be the control subjects of a feedback control loop. Typically, there will be five to six control points at the macro process level for variables associated with external output, external input, key intermediate products, and other high-leverage process points. Control points in the special-contract management process are represented graphically in Figure 8.4. Feedback loop design and other issues surrounding process control are covered in detail in Chapter 7.

Analyzing the Process

Process analysis is performed for the following purposes:

- Assess the current process for its effectiveness and efficiency.
- Identify the underlying causes of any performance inadequacy.
- Identify opportunities for improvement.
- Make the improvements.

First, referring to the process flowchart, the team breaks the process into its component activities using a procedure called *process decomposition*, which consists of progressively breaking apart the process, level by level, starting at the macro level. As decomposition proceeds, the process is described in ever-finer detail.

As the strengths and weaknesses of the process are understood at one level, the BPM team's interim theories and conclusions will help decide where to go next with the analysis. The team will discover that certain subprocesses have greater influence on the performance of the overall business process than others (an example of the Pareto principle). These more significant subprocesses become the target for the next level of analysis.

Decomposition is complete when the process parts are small enough to judge as to their effectiveness and efficiency. Table 8.2 shows examples from three levels of decomposition (subprocess, activity, and task) of three typical business processes (procurement, development engineering, and office administration).

Table 8.2 Three Levels of Decomposition

Business Process	Subprocess	Activity	Task
Procurement	Supplier selection	Supplier survey	Documentation of outside supplier
Development engineering	Hardware design	Engineering change	Convening the change board
Office administration	Providing administrative support services	Managing calendars	Making a change to existing calendar

Measurement data are collected according to the measurement plan to determine process effectiveness and efficiency. The data are analyzed for effectiveness (conformance to customer needs) and long-term capability to meet current and future customer requirements.

The goal for process efficiency is that all key business processes operate at minimum total process cost and cycle time while still meeting customer requirements.

Process *effectiveness* and *efficiency* are analyzed concurrently. Maximizing effectiveness and efficiency together means that the process produces high quality at low cost; in other words, it can provide the greatest *value* to the customer.

Business process adaptability is the ability of a process to readily accommodate changes in both the requirements and the environment while maintaining its effectiveness and efficiency over time. To analyze the business process, the flow diagram is examined in four steps and modified as necessary.

The Process Analysis Summary Report is the culmination and key output of this process analysis step. It includes the findings from the analysis, that is, the reasons for inadequate process performance and potential solutions that have been proposed and recorded by owner and team as analysis progressed. The completion of this report is an opportune time for an executive owner/stakeholder review.

The owner/stakeholder reviews can be highly motivational to owners, teams, and stakeholders. Of particular interest is the presentation of potential solutions for improved process operation. These have been collected throughout the planning phase and stored in an idea bin. The design suggestions are now documented and organized for executive review as part of the Process Analysis Summary Report presentation.

In reviewing potential solutions, the executive owner and the quality council provide the selection criteria for acceptable process design alternatives. Knowing upper management's criteria for proposed solutions helps to focus the process management team's design efforts and makes a favorable reception for the reengineered new process plan more likely.

Redesigning the Process

In process design, the team defines the specific operational means for meeting stated product goals. The result is a newly developed process plan.

Design changes fall into five broad categories: work flow, technology, people and organization, physical infrastructure, and policy and regulations.

In the design step, the owner and the team must decide whether to create a new process design or to redesign the existing process. Creating a new design might mean radical change; redesign generally means incremental change with some carryover of existing design features.

The team will generate many design alternatives, with input from both internal and external sources. One approach to generating these design alternatives from internal sources is to train task-level performers to apply creative thinking to the redesign of their process.

Ideas generated in these sessions are documented and added to the idea bin. Benchmarking can provide a rich source of ideas from external sources, including ideas for radical change. Benchmarking is discussed in detail in Chapter 9.

In designing for process effectiveness, the variable of greatest interest is usually process cycle time. In service-oriented competition, lowest process cycle time is often the decisive feature. Furthermore, cycle-time reduction usually translates to efficiency gains as well. For many processes, the most promising source of cycle-time reduction is the introduction of new technology, especially information technology.

Designing for speed creates surprising competitive benefits: growth of market share and reduction of inventory requirements. Hewlett-Packard, Brunswick Corp., GE's Electrical Distribution and Control Division, AT&T, and Benetton are among the companies that have reported stunning achievements in cycle-time reduction for both product development and manufacturing (Dumaine, 1989). In each of the companies, the gains resulted from efforts based on a focus on major processes. Other common features of these efforts included the following:

- Stretching objectives proposed by top management
- Absolutely adhering to the schedule, once agreed to
- Applying state-of-the-art information technology
- Reducing management levels in favor of empowered employees and self-directed work teams
- Putting speed in the culture

In designing for speed, successful redesigns frequently originate from a few relatively simple guidelines: eliminate handoffs in the process, eliminate problems caused upstream of activity, remove delays or errors during handoffs between functional areas, and combine steps that span businesses or functions. A few illustrations follow:

- *Eliminate handoffs in the process.* A *handoff* is a transfer of material or information from one person to another, especially across departmental boundaries. In any process involving more than a single person, handoffs are inevitable. It must be recognized, however, that the handoff is time-consuming and full of peril for process integrity—the missed instruction, the confused part identification, the obsolete specification, the miscommunicated customer request. In the special-contract management process, discussed earlier in this chapter, the use of concurrent review boards eliminated the 28 sequential executive approvals and associated handoffs.
- *Eliminate problems caused upstream of activity.* Errors in order entry at a U.S. computer organization were caused when sales representatives configured systems incorrectly. As a result, the cost of the sales-and-order process was 30 percent higher than that of competitors, and the error rates for some products were as high as 100 percent. The cross-functional redesign fixed both the configurations problem and sales force skills so that on-time delivery improved at significant cost savings (Hall et al., 1993).
- *Remove delays or errors during handoffs between functional areas.* The processing of a new policy at a U.K. insurance organization involved ten handoffs and took at least 40 days to complete. The organization implemented a case manager approach by which only one handoff occurred, and the policy was processed in less than seven days (Hall et al., 1993).
- *Combine steps that span businesses or functions.* At a U.S. electronics equipment manufacturer, as many as seven job titles in three different functions were involved in the nine steps required to design, produce, install, and maintain hardware. The organization eliminated all but two job titles, leaving one job in sales and one job in manufacturing (Hall et al., 1993).

Process design testing is performed to determine whether the process design alternative will work under operating conditions. Design testing may include trials, pilots, dry runs, simulations, etc. The results are used to predict new process performance and cost/benefit feasibility.

Successful process design requires employee participation and involvement. To overlook such participation creates a lost opportunity and a barrier to significant improvement. The creativity of the first-line workforce in generating new designs can be significant.

Creating the New Process Plan

After we have redefined a key process, we must document the new process and carefully explain the new steps. The new process plan now includes the new process design and its control plan for maintaining the new level of process performance. The new process plan for the special-contract management process, shown as a high-level process schematic, is shown in Figure 8.5.

The Transfer Phase: Transferring the New Process Plan to Operations

There are three steps in the transfer phase: (1) planning for implementation problems, (2) planning for implementation action, and (3) deploying the new process plan.

Planning for Implementation Problems

A major BPM effort may involve huge expenditures and precipitate fundamental change in an organization, affecting thousands of jobs. All this poses major management challenges. All the many changes must be planned, scheduled, and completed so that the new process may be deployed to operational management. Table 8.3 identifies specific categories of problems to be addressed and the key elements that are included.

Of the five categories listed in Table 8.3, *people and organization* is usually the source of the most challenging change issues in any BPM effort. Implementation issues in the people and organizational design category include new jobs, which are usually bigger; new job descriptions; training people in the new jobs; new performance plans and objectives; new

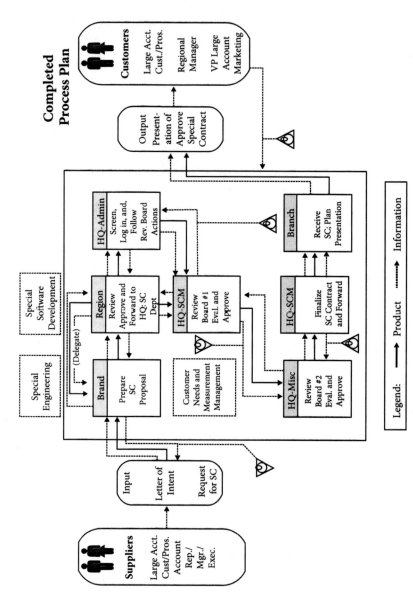

Figure 8.5 High-level process schematic.

Table 8.3 Specific Categories of Problems to be Addressed

Category	Key Elements
Workflow	Process anatomy (macro/micro; cross-functional; intra-functional; inter-departmental; intra-departmental)
Technology	Information technology; automation
People and organization	Jobs; job description; training and development, performance management; compensation (incentive-based or not); recognition/reward; union involvement; teams; self-directed work teams; reporting relationships; de-layering.
Infrastructure (physical)	Location; space; layout; equipment; tools; furnishings
Policy/regulations	Government; community; industry; company; standards; culture
New-process design issues	Environmental, quality, costs, sourcing

compensation systems (incentive pay, gain sharing, and the like); new recognition and reward mechanisms; new labor contracts with unions; introduction of teamwork and team-building concepts essential to a process orientation; formation of self-directed work teams; team education; reduction of management layers; new reporting relationships; development and management of severance plans for those whose jobs are eliminated; temporary continuation of benefits; outplacement programs; and new career paths based on knowledge and contribution, rather than on promotion within a hierarchy. The list goes on. Additionally, there are changes in technology, policy, and physical infrastructure to deal with.

The importance of change management skills becomes clear. Deploying a new process can be a threat to those affected. The owner and the team must be skilled in overcoming resistance to change.

Creating Readiness for Change

Change happens when four conditions are combined. First, the current state must be seen as unsatisfactory, even painful; it must constitute a tension for change. Second, there must be a satisfactory alternative, a vision of how things can be better. Third, some practical steps must be available to reach the satisfactory state, including instruction in how to take the steps and support during the journey. Fourth, to maintain the change, the organization and individuals must acquire skills and reach a state of self-efficacy.

These four conditions reinforce the intent to change. Progress toward that change must be monitored continuously to make the change permanent. In the operational management phase, operational controls, continuous improvement activity, and ongoing review and assessment all contribute to ensuring that the new process plan will continue to perform as envisioned.

Planning for Implementation Action

The output of this step is a complex work plan, to be carried out by the owner and the BPM team. They will benefit from skills in the techniques of project management.

Deploying the New Process Plan

Before actually implementing the new process, the team tests the process plan. They test selected components of the process and may carry out computer simulations to predict the performance of the new process and determine its feasibility. Also, tests help the team refine the *rollout* of the process and decide whether to conduct parallel operation (old process and new process running concurrently). The team must decide how to deploy the new process. There are several options:

- ▲ Horizontal deployment, function by function.
- ▲ Vertical deployment, top down, all functions at once.
- ▲ Modularized deployment, activity by activity, until all are deployed.
- ▲ Priority deployment, subprocesses and activities in priority sequence, those having the highest potential for improvement going first.
- ▲ Trial deployment, a small-scale pilot of the entire process, then expansion for complete implementation. This technique was used in the first redesign of the special-contract management process; that is, a regional trial preceded national expansion. The insurance organization USAA conducts all pilot tests of new process designs in their Great Lakes region. In addition to "working the bugs out of the new design before going national," USAA uses this approach as a "career-broadening

Process Plan

Process purpose or mission
Process goals and targets
Process management infrastructure (process owner, team, stakeholder)
Process contract
Process description and model
Customer requirements (customer list, customer needs, requirements statement)
Process flow
Measurement plan
Process analysis summary report
Control plan
Implementation action plan
Resource plan
Schedules and timeline

Figure 8.6 Contents of a new process plan.

experience for promising managers" and to "roll out the new design to the rest of the organization with much less resistance" (Garvin, 1995).

Full deployment of the new process includes developing and deploying an updated control plan. Figure 8.6 lists the contents of a new process plan.

Operational Management Phase: Managing the New Process

The operational management phase begins when the process is put into operation. The major activities in operational management are (1) process quality control, (2) process quality improvement, and (3) periodic process review and assessment.

Business Process Metrics and Control

Process control is an ongoing managerial process, in which the actual performance of the operating process is evaluated by measurements taken at the control points, comparing the measurements to the quality targets, and taking action on the difference. The goal of process control is to maintain performance of the business process at its planned level.

Business Process Improvement

By monitoring process performance with respect to customer requirements, the process owner can identify gaps between what the process is delivering and what is required for full customer satisfaction. These gaps are targets for process quality improvement efforts. They are signaled by defects, complaints, high costs of poor quality, and other deficiencies. (See Chapter 7, "Assuring Repeatable and Compliant Processes.")

Periodic Process Review and Assessment

The owner conducts reviews and assessments of current process performance to ensure that the process is performing according to plan. The review should include a review and an assessment of the process design itself to protect against changes in the design assumptions and anticipated future changes, such as changes in customer needs, new technology, or competitive process designs. It is worthwhile for the process owner to establish a schedule for reviewing the needs of customers and evaluating and benchmarking the present process.

BPM is a natural extension of many of the lessons learned in early quality improvement activities. It requires a conceptual change—from reliance on functional specialization to an understanding of the advantages of focusing on major business processes. It also requires an additional piece of organization machinery: an infrastructure for each of the major processes.

The Future of BPM Combined with Technology

Business process management is being combined with service-oriented architectures (SOAs), technologies, and performance excellence tools such as Lean and Six Sigma to accelerate improvements and results. At the same time, these tools are increasing organizational flexibility and technology-enabled responsiveness. Many successful organizations have found that the linkages are clear.

According to IBM (2013), early adopters who have worked their way past cultural and organizational barriers are seeing such impressive performance and financial results as these:

- Improved responsiveness to market challenges and changes through aligned and significantly more flexible business and technical architectures.
- Improved ability to innovate and achieve strategic differentiation by driving change into the market and tuning processes to meet the specific needs of key market segments.
- Reduced process costs through automation and an improved ability to monitor, detect, and respond to problems by using real-time data, automated alerts, and planned escalation.
- Significantly lower technical implementation costs through shared process models and higher levels of component reuse.
- Lower analysis costs and reduced risk through process simulation capabilities and an improved ability to gain feedback and buy-in prior to coding.

The rewards can be great, especially for those who take action now.

References

Dumaine, B. (1989). "How Managers Can Succeed Through Speed." *Fortune*, Feb. 13, pp. 54–60.

Garvin, D. A. (1995). "Leveraging Processes for Strategic Advantage." *Harvard Business Review*, September/October, vol. 73, no. 5, pp. 77–90.

Hall, G., J. Rosenthal, and J. Wade. (1993). "How to Make Reengineering Really Work." *Harvard Business Review*, November/December, vol. 71, no. 6, pp. 107–128.

Hardaker, M. and B. Ward. (1987). "Getting Things Done: How to Make a Team Work," *Harvard Business Review*, November/December., vol. 65, pp. 112–119.

Juran J. M. and DeFeo J. A. (2010). *Juran's Quality Handbook: The Complete Guide to Performance Excellence*, 6th ed. McGraw-Hill, New York.

Kaplan, R. S., and D. P. Norton. (1992). "The Balanced Scorecard—Measures that Drive Performance." *Harvard Business Review,* January/February, vol. 7, no. 1, pp. 71–79, reprint #92105.

Pall, G. A. (1987). *Quality Business Process Management.* Prentice-Hall, Inc., Englewood Cliffs, New Jersey.

Riley, J. F., Jr. (1989). *Executive Quality Focus: Discussion Leader's Guide.* Science Research Associates, Inc., Chicago.

Riley, J. F., Jr., G. A. Pall, and R. W. Harshbarger. (1994). *Reengineering Processes for Competitive Advantage: Business Process Quality Management (BPQM),* 2nd ed. Juran Institute, Inc., Wilton, Conn.

Skalle, H. and B. Hahn. (2013). "Applying Lean, Six Sigma, BPM, and SOA to Drive Business Results". *Redguides for Business Leaders.* IBM Corporation, Armonk, New York.

CHAPTER 9

Benchmarking to Sustain Market Leadership

This chapter defines the use of benchmarking as an effective tool to sustain organizational performance and aid in strategic and annual goal setting.

Benchmarking: What It Is and What It Is Not

Benchmarking has been in existence for a great many years. The concept of one individual observing how another performs a given task and then applying any learning from that to adapt and improve how the task is executed is one of the fundamental ways in which human beings learn and develop. In the context of business, learning from one's competitors has also been in existence for as long as business has. However, the application of learning from best practices to the business environment in a structured, methodical, and indeed legal and ethical way is relatively new. Xerox Corp. is most commonly credited with developing the modern form of benchmarking, and it is fair to say that the majority of today's benchmarking practices are built upon the approach developed by them in the 1970s.

Although their story has previously been well told in a multitude of management texts, it is still worthy of brief comment here to set the scene. A combination of poor product quality, high overheads, and increasing competition from a growing number of Japanese organizations had left Xerox in a precarious position in the late 1970s. A visit to Japan provided the wake-up call that change was essential if they were to survive (Kearns and Nadler, 1993). They put in place a series of benchmarking activities aimed at identifying the best-performing organizations in various aspects of their business and determining what these organizations were doing that enabled a superior performance. Most famous is the benchmarking of logistics operations that they undertook with L.L. Bean (Camp, 1989). With this, modern benchmarking was born.

Benchmarking has evolved to become an essential element of the business performance improvement tool kit and is now frequently used by many organizations in a wide range of different industries. But despite this, it remains one of the most widely misunderstood improvement tools. It means many different things to many different people, and all too frequently benchmarking projects fail to deliver on their promise of improvement or real results.

Executed correctly, benchmarking will provide a powerful focus for organizations, driving home the facts and convincing the organization of the need to embark upon improvement strategies. Benchmarking is a tool that enables the identification and ultimately the achievement of excellence, based upon the realities of the business environment rather than on internal standards and historical trends.

Benchmarking is not what we would term *industrial tourism* in which superficial industrial visits are undertaken in the absence of any point of reference and do not assist in the improvement process. It is impossible to acquire detailed knowledge after only a quick glance or one short visit, and it is rare for such visits to result in an action plan that will lead to improvement. In the absence of prior benchmarking, it is also difficult to identify which organizations should be visited, and so there is a real risk that visits are made to organizations that are perceived as being the best or at least better, when the reality may be very different. However, there is a valuable role to be played by this type of site visit, when it is conducted following a structured benchmarking analysis and the organization being visited has been identified as a best performer.

Benchmarking also should not be considered a personal performance appraisal tool. The focus should be on the organization and the individuals within it. Failure to adopt this philosophy will only lead to resistance and will undoubtedly add roadblocks to a successful benchmarking journey.

Nor should benchmarking be a momentary glimpse, but rather it should be considered a continuous process. Organizations must change performance rapidly to remain competitive in business environments today. This fast-paced tempo is further accelerated in sectors where benchmarking is commonplace, where businesses rapidly and continuously learn from one another. A prime example comes from the oil and gas industry where organizations have to respond to ever-increasing business, technological, and regulatory demands. The majority of the key players in this industry are participating in focused benchmarking consortia on an annual basis. It is also much more than a competitive analysis. Benchmarking goes further than examining the pricing and features of competitors' products or services. It considers not only the output but also the process by which the output was obtained. Benchmarking is also much more than market research, considering the business practices in place that are enabling the satisfaction of customer needs and thus realizing superior business performance. It provides evidence-based input offering a powerful focus for management, driving home the facts and convincing the organization of the need to embark on improvement activities.

Participating in benchmarking should also not be viewed as a standalone activity. To succeed, it must be part of a continuous improvement strategy, it must be conducted regularly, and it should be enveloped in the continuous improvement culture of an organization. Like any other project, it has to have the full support of senior management, the resources necessary to fulfill the objectives, and a robust project plan that is adhered to.

Finally, benchmarking should not be viewed as the answer in itself. It is a means to an end. An organization will not improve performance by benchmarking alone. It must act upon the findings of the benchmarking to improve. The output of benchmarking should provide input to decision making or improvement action planning. This requires detailed consideration of the benchmarking analysis, formulation of learning points, and development of action plans in order to implement change and realize improvements.

So how can we define benchmarking? A scan of the literature will quickly reveal myriad definitions (Anand and Kodali, 2008), each offering a slight variation on a common theme. Rather than repeat these here, we prefer to offer our own definition:

> *Benchmarking* is a systematic and continuous process that facilitates the measurement and comparison of performance and the identification of best practices that enable superior performance.

This definition is deliberately generic so that it can encompass all types of benchmarking. In this context, measurement and comparison may be between organizations, business units, business functions, and business processes, products, or services. The benchmarking may be internal or external, between competitors, within the same industry, or cross-industry. Regardless of the category of benchmarking, this definition still applies.

Objectives of Benchmarking

The objectives of benchmarking can be summarized as follows:

- Determine superior performance levels.
- Quantify any performance gaps.
- Identify best practices.
- Evaluate reasons for superior performance.
- Understand performance gaps in key business areas.
- Share knowledge of working practices that enable superior performance.
- Enable learning to build foundations for performance improvement.

When one is talking of superior performance, ultimately of course the aim should be to achieve world-class performance. However, in reality it is often difficult to be able to ensure that the world's leading performers are participating in a given benchmarking exercise. Instead, benchmarking partners should be selected carefully to ensure the output will provide the required added value.

Once superior performance has been determined, the gap between this and the performance level of the benchmarker is quantified. The working practices enabling superior performance are identified and the enablers evaluated. This knowledge is then shared among benchmarkers to enable the learning to be taken away and implemented as part of a performance improvement program.

Thus benchmarking can be viewed as a two-phase process, where phase 1 is a positioning analysis aimed at identifying gaps in performance and phase 2 is focused upon learning from those best practices that enable superior performance.

Why Benchmark?

There are two good reasons for organizations to benchmark themselves. First, it will help them stay in business by offering opportunities to become better than other similar organizations, competitors or not. Second, it ensures that an organization is continually striving to improve its performance through learning. Benchmarking opens minds to new ideas from sources either within the same industry or from many other unrelated industries, identifying how those who have demonstrated performance leadership work.

Yet many organizations benchmark simply to be able to demonstrate to stakeholders, be they customers, shareholders, lenders, regulators, etc., that the organization is performing at an acceptable level. Of course this is a perfectly legitimate reason for benchmarking, although the real potential value of the technique is missed by narrowing the focus in this way.

Benchmarking also provides a very effective input to an organization's strategic planning processes by establishing credible goals and realistic targets based upon external references.

To really grasp the intent of benchmarking, an organization should be benchmarking not only to demonstrate good performance but also to identify ways in which it can change its practice to significantly improve its performance. Those organizations with a strong performance improvement culture will be benchmarking continuously as this provides them with objective evidence of where to focus improvement activities, how much they should be improving, and what changes to their working practices they might consider to realize improvements.

Classifying Benchmarking

There are many different ways to classify benchmarking (Table 9.1), and the literature is full of different classifications (Anand and Kodali, 2008) that make it very confusing for someone new to the topic to really understand what benchmarking is and which approach is best for her or him. The fact of the matter is that there is an underlying process that can be considered generic to almost all types of benchmarking. However, to provide some clarity on the differences in classification, we have considered benchmarking in terms of what is to be benchmarked, who the benchmarking is going to involve, and how the benchmarking is to be conducted:

- Subject matter and scope (what)
- Internal and external, competitive and noncompetitive benchmarking (who)
- Data and information sources (how)

Table 9.1 Ways in Which Benchmarking Is Often Classified

Classification Criteria		
Subject Matter (What)	**Participants (Who)**	**Data Sources (How)**
Functional benchmarking	Internal benchmarking	Database benchmarking
Process benchmarking	External benchmarking	Survey benchmarking
Business unit or site (location) benchmarking	Competitive benchmarking	Self-assessment benchmarking
Projects benchmarking	Noncompetitive benchmarking (same industry and cross-industry)	One-to-one benchmarking
Generic benchmarking		Consortium benchmarking
Business excellence models		

Source: Juran Institute, Inc., 2013.

Subject Matter and Scope (What)

Benchmarking is often categorized according to what is being benchmarked. Typical categories include:

- Functional benchmarking
- Process benchmarking
- Business unit or site (location) benchmarking
- Project benchmarking
- Generic benchmarking
- Business excellence models

Functional Benchmarking

Functional benchmarking describes the process whereby a specific business function forms the focus for the benchmarking. In the context of the organization, this may involve benchmarking several different business units or site locations. Typical examples of functional benchmarking include the analysis of the procurement, finance, Internet technology (IT), safety, operations, or maintenance functions. The analysis focuses upon all aspects of the function rather than on the processes involved and the specific activities conducted.

Process Benchmarking

In process benchmarking the focus of the study is upon a specific business process or a part thereof. Examples include product development, invoicing, order fulfillment, contractor management, and customer satisfaction management. Process benchmarking will often involve several functional groups and may also involve many different site locations. There is often a lot of overlap between what is termed *functional* benchmarking and *process* benchmarking (e.g., a benchmarking of the procurement process may look very similar to a benchmarking of the procurement function). Many business processes are not specific to any one industry and so can benefit from broadening participation in the analysis to organizations from a multitude of industries.

Business Unit or Site (Location) Benchmarking

Benchmarking individual business units or site locations against one another is often (but not always) seen in internal benchmarking studies within a single organization. The performance of each unit is analyzed and compared to that of other units. This analysis may incorporate all activities of each unit in their entirety or may be confined to selected functional groups or business processes. For example, Juran manages an annual benchmarking consortium comparing the performance of many of the world's oil and gas processing facilities. Each of the key business processes is included in the analysis, and participants come from a wide range of different organizations.

Project Benchmarking

This type of benchmarking focuses upon projects undertaken by organizations. Because projects vary widely in their nature, these studies are normally tailored for specific project types. For example, one may benchmark oil pipeline construction projects, software implementation projects, facility decommissioning projects, etc. Normally included are all the business processes pertaining to the project being analyzed, although the scope may often be limited to a subset of processes. For example, a construction project benchmarking may focus specifically upon contractor selection, procurement, and commissioning.

Generic Benchmarking

Generic benchmarking considers all business processes required to achieve a certain level of performance in a given area. The focus is upon the result and what is required to achieve it. For example, a hospital may undertake a generic benchmarking exercise to identify ways in which it can reduce treatment waiting times. In so doing it may benchmark across a number of different industries where customer waiting times are of paramount importance, e.g., insurance claims processing, vessel clearance procedures for major waterways (e.g., Suez Canal), and calamity response times for the different emergency services (police, fire, ambulance). Inevitably, there is a lot of overlap between process benchmarking and generic benchmarking although in the latter there is often less emphasis on gap analysis and greater emphasis upon a detailed consideration of working practices.

Business Excellence Models

Business excellence models have been developed to provide a framework by which organizations can holistically measure and therefore improve their performance. The purpose of their design is such that they encompass all key aspects of an organization that drive performance.

Although these models are designed to support a self-assessment process, they also lend themselves to providing an excellent framework for comparative benchmarking, although they are infrequently used for this purpose. Benchmarking in this way using such models is essentially a form of generic benchmarking whereby all elements required for excellence are considered. Furthermore, a requirement of both the Baldrige and European Foundation for Quality Management (EFQM) models is that organizations be able to demonstrate benchmarking activity.

Internal and External, Competitive and Noncompetitive Benchmarking (Who)

Benchmarking studies are frequently classified by type of participant. Depending on the type of benchmarking being undertaken, it is not always possible to have any control over participant selection. But where this is possible, the selection of others to benchmark with is one of the first tasks and often the most difficult one at the outset of any benchmarking study. Potential participants will be identified according to a range of criteria, the main one being the perceived performance level (where superior or world-class performance is the aim).

The four main types of benchmarking are:

1. Internal benchmarking
2. External benchmarking
3. Competitive benchmarking
4. Noncompetitive benchmarking

Each of these has specific benefits and drawbacks that need to be considered when selecting the most suitable benchmarking approach. These include:

- The similarity between the participants in terms of the subjects to be benchmarked
- The level of control over the benchmarking process
- The cost and time input required to conduct the benchmarking
- The degree of openness that is possible and the level of confidentiality necessary
- The potential for learning and therefore performance improvement

Internal Benchmarking

Internal benchmarking is the comparison of performance and practices of similar operations within the same organization. Depending upon the size of the organization and the nature of its business, this may or may not be feasible, for the organization would need to have duplicate groups conducting the same activities. Should this be the case, internal benchmarking is often a popular first step as it allows organizations to prepare themselves for broader benchmarking activities within the safety of their own environment where they have full control over the process. This is likely also to be the least costly and time-consuming way to benchmark. But the potential for finding performance leaders is much smaller, and the opportunity for learning is usually more limited.

External Benchmarking

External benchmarking involves participants from different organizations. The opportunity for learning is normally greater than that achievable by internal benchmarking, but there is obviously a requirement to share information outside of the organization. This brings with it some potential restraints. There will almost certainly be limits on the data that organizations are willing to share, especially if the other participants are competitors, and there will of course be a need for stricter confidentiality. External benchmarking is further categorized according to the nature of the participants who can be competitors (competitive benchmarking) or not (noncompetitive benchmarking). These are considered below.

Competitive Benchmarking

Competitive benchmarking is a form of external benchmarking in which the participants are all in competition with one another. By definition, the participants in a competitive benchmarking program are from the same industry, and the focus is normally upon industry-specific processes. For example, Juran has studied patient safety performance among different hospitals. This normally brings with it a high degree of sensitivity which needs to be carefully managed for a successful outcome to be realized; but when it is conducted properly, the results can be very valuable. Conversely, topics that are not directly related to the core business of the competing organizations are usually less sensitive to benchmark among competitors. But ironically these more generic topics are not normally those that an organization wishes to benchmark with its competitors as greater value is more frequently gained from benchmarking such topics cross-industry. There are also likely to be some subject areas that most organizations will not be willing to benchmark with competitors, such as proprietary processes or products and innovations that provide competitive advantage.

Noncompetitive Benchmarking

Noncompetitive benchmarking is a form of external benchmarking in which the participants are not in direct competition with one another. They may be from within the same industry or cross-industry. For example, an organization operating a container port in the United States may benchmark with another in Europe. Although they are in the same industry, they are unlikely to be competitors as they are operating in different markets. Juran manages an annual global benchmarking consortium for gas pipelines. The participants are all in the same industry and are primarily interested in benchmarking those processes specific to their industry. But they are not in direct competition with one another as they operate in totally different marketplaces, delineated by geographical region. This means that they are very willing to share knowledge and practices openly for mutual benefit without fear of giving anything away that may impact their competitiveness.

In cross-industry noncompetitive benchmarking, those subject areas that are not industry-specific are most commonly analyzed, and it is in this classification that most generic benchmarking studies sit. These tend to be support processes such as administration, human resources, R&D, finance, procurement, IT, and health, safety, and environment (HSE). Cross-industry external benchmarking potentially offers the greatest opportunities for learning and performance improvement for a number of reasons. First, the pool of potential participants is much bigger. Second, participants thought to be superior performers in the subject area being benchmarked can be identified and invited to participate. Third, the willingness to share knowledge will be greatest where there is no fear of competitive sensitivity.

Data and Information Sources (How)

Benchmarking can also be classified according to the source of the data used in the comparative analysis. Such classification can be made in many ways, but the list below addresses the main categories found:

- Database benchmarking
- Survey benchmarking
- Self-assessment benchmarking

- One-to-one benchmarking
- Consortium benchmarking

Database Benchmarking

In this type of benchmarking, data from a participant are compared to an existing database containing performance data. An analysis is performed, and the results are provided to the participant. Benchmarking in this way normally requires a third party to administer the database and produce the analyses. The development of the Internet in recent years has led to the growth in this type of benchmarking as it can be easily administered online. The participating organization can submit its data via an online questionnaire and receive a report of the analyses online, usually in a very short time. A quick search of the Internet will reveal the large number and wide range of online benchmarking databases available.

This type of benchmarking is also sometimes offered by consultancy organizations that have accumulated performance data pertaining to specific activities. For example, Juran has been benchmarking in the oil and gas industry since 1995 and during this time has developed a comprehensive database of performance figures relating to this industry. And because the data have been well defined and thoroughly validated during the collection process they are extremely reliable. It is therefore an excellent data source against which oil and gas organizations can be benchmarked.

Many organizations start out on their benchmarking journey by purchasing data from proprietors of such databases. Although this type of benchmarking can be very useful in providing fast feedback on performance, it can have drawbacks. The participant has no control over the content of the analysis and has to accept the metrics that are used to determine performance. Often the source of the data is not disclosed; thus it may be difficult for the benchmarker to assure himself or herself of its relevance. The metrics used may not be clearly defined and may not be validated effectively, resulting in poor data quality and flawed analyses. Care should therefore be taken when entering into this type of benchmarking, and the participant should realize the potential shortcomings. For best results, only a bona fide consultant with a sterling reputation and a good track record should be sought.

Survey Benchmarking

This term is used to describe benchmarking exercises conducted via the completion of a survey or a review process. Typically, a survey document is sent to participating organizations to be completed and returned. Sometimes the survey documents are sent to organizations without their prior agreement to participate, in the hope that they will complete the survey and return it. Of course, this approach is nearly always less successful, with a relatively poor return rate.

The survey may be organized by a third-party consultant or by one of the participating organizations, although in the latter case there will be greater restrictions on what data can be shared directly among participants to ensure compliance with antitrust legislation. Sometimes there may be a fee involved for organizations to participate, and sometimes a single organization or even a consultancy may sponsor the entire exercise, in which case the output for the other participants is often less sophisticated.

The potential drawbacks of this approach are similar to those encountered with database benchmarking in that each organization has minimal control over the benchmarking process, the metrics may not be defined adequately, and the validation of submitted data may be limited. Nonetheless, this type of approach can provide a useful, albeit limited, comparative analysis with limited effort required on the part of each participating organization.

The survey process can be extended to include a review element, whereby the benchmarking coordinator (normally a third-party consultant) will visit each of the participating organizations as part of the survey process. This allows the consultant to delve more deeply into specific areas to gain richer data (often qualitative data), which can better inform the learning process of the participants, particularly in the area of assessing working practices that underpin superior performance.

Self-Assessment Benchmarking

As previously discussed, self-assessment is an integral part of many performance excellence models. These self-assessments can be used for generic benchmarking among organizations across all industries. These models provide an excellent framework for comparative benchmarking whereby all elements required for excellence are considered. The analysis will often focus on not just quantitative data analysis but also a qualitative view of working practices. However, there is an inherent weakness in the process associated with the subjective nature of self-assessment. Sometimes third parties (consultants) are employed to oversee the process to introduce some level of objectivity or even to conduct the assessments.

One-to-One Benchmarking

This type of benchmarking is probably most commonly reported in the literature, but as we pointed out, benchmarking is not industrial tourism, whereby relatively superficial site visits are conducted among organizations to explore performance. Such exchanges rarely bring fruitful insights, and there is always the uncertainty of whether the organization being visited is really a superior performer.

However, if after conducting a benchmarking study, an organization is identified as delivering superior performance levels in the subject area of interest, then a one-to-one benchmarking can investigate specific areas in much greater depth and deliver rich information pertaining to the drivers of superior performance. This approach is common in consortium benchmarking where, having received the benchmarking analysis, two organizations will agree to benchmark further one to one to obtain more detailed understanding in specific working practices. A good example comes from a cross-industry procurement benchmarking consortium managed by Juran. Following participation in the study, two of the participants agreed to undertake a one-to-one benchmarking. One organization was particularly strong in contract tendering and contractor selection, while the other demonstrated superiority in strategic procurement. Cooperating in this way led to a greater understanding of leading working practices in each of these areas and improved performance for both parties.

Consortium Benchmarking

Without doubt, this form of benchmarking has the greatest potential to deliver improved performance for its participants. A consortium is formed of participants, usually (but not always) supported by a third-party facilitator. They agree on the participants to be invited; the subjects to be benchmarked; the methodology to be followed; the metrics (and their definitions) to be used; the validation criteria; the nature of the analysis, reporting, and deliverables; and the time scales to be adhered to. Thus the participants have a very high level of control over the entire process, and the outcomes from the process are normally reliable data, thorough analysis, and valuable results. This approach does require a great effort on behalf of each participant to achieve the desired outcome. It is therefore more time-consuming and often more costly to undertake, but the added value is normally far in excess of that achieved through other benchmarking approaches.

We have demonstrated that benchmarking can be classified in many different ways according to the subject matter of the analysis, the nature of the benchmarking participants, data sources, and the methodologies employed. However, differentiating in this way is largely academic, and although these different approaches have their inherent pros and cons and some are clearly more effective than others, they all should have the same ultimate objective—to provide learning on how to improve business performance.

Benchmarking and Designing New Products

The main objective of designing for quality is to prepare organizations so that they are able to meet their performance goals. In so doing they must do the following:

1. Identify customers.
2. Determine customer needs.
3. Develop product (service) features required to meet customer needs.
4. Establish quality goals that meet customer needs.
5. Develop processes that deliver the product (service) features.
6. Prove that the organization's processes can meet the goals.

Benchmarking can offer input to this process by providing the vehicle whereby organizations can learn from best practices and incorporate that learning into designing new and improved business processes. It also enables performance planning goals to be established based upon the reality of what is achievable by other benchmarked organizations. Hidden customer needs may be revealed by the benchmarking and product or service features identified.

Benchmarking and Long- and Short-Term Planning

Long-term plans or key strategies derived from the vision will comprise strategic goals addressing all aspects of the organization's performance including business process performance, product or service performance, customer satisfaction, the cost of poor quality, and the organization's competitive performance. By necessity, these strategic goals will be constantly evolving. Benchmarking analyses enable an organization to set these goals on the external reality and ensure it focuses on closing the gaps between actual and envisioned performance.

The findings from benchmarking enable organizations to understand exactly how much improvement is required for attainment of superior performance. Frequent and regular benchmarking supports the establishment of specific and measurable short-term plans, based upon reality rather than historical performance, resulting in step-by-step improvements in performance over time (Figure 9.1). The objective is for the organization to overtake the performance leaders, turning a performance gap into superior performance leadership.

An implementation process is required to convert the long- and short-term plans into operational plans. This requires organizations to determine

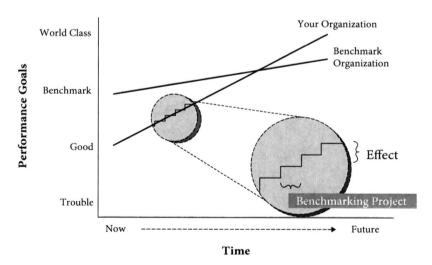

Figure 9.1   Benchmarking over time. (From Juran Institute, Inc., 2013)

exactly how their specific strategic goals are to be met, identify the actions required to enable this, determine who has the responsibility for carrying out the actions, calculate and allocate the resources required, and plan, schedule, and control the implementation. The output from benchmarking once again provides the external perspective and feeds into this process by offering information related to the best practices that have been identified.

Organizations should review their performance on a regular basis to determine progress against the goals set and to measure the gap between the current state and the vision. Benchmarking is the perfect vehicle to support this review process by providing objective evidence of current performance, determining the gaps in performance levels being achieved by other organizations, identifying best practices, and offering the opportunity to learn from the leading performers.

Thus it is clear that benchmarking is a powerful tool that can contribute significantly to an organization's ability to effectively and strategically manage its performance. It forces organizations and their managers to consider the broader perspective, to look outside of their comfort zones, to learn from those identified as excellent performers; and it fuels the drive for change. By revealing what the best-performing organizations are already achieving and by establishing a factual base for best practices, benchmarking enables organizations to manage their performance to achieve world-class leadership.

The Benchmarking Process

Critical to the success of any benchmarking program are a number of key factors:

- Scope out the study and determine objectives.
- Identify and define all metrics.
- Agree on a schedule and stick to it.
- Ensure resources are available to support the benchmarking.
- Provide support to participants throughout the process.
- Validate all data.
- Normalize the data.
- Clearly and effectively report the findings.
- Enable sharing of best practices.

Irrespective of the type of benchmarking undertaken, it is essential that a well-structured and systematic process be followed to realize these critical success factors. There are many such benchmarking processes described in the literature (Anand and Kodali, 2008), but the pioneering model from which most others have been formulated is that used by Xerox and described by Camp (1989). Camp's 10-step benchmarking process was also described in the fifth edition of this handbook (Camp and DeToro, 1999). Since that time the Juran Institute has published its own 7-Step Benchmarking Process, which was developed over a period of many years and has formed the basis of a multitude of annual benchmarking consortia since 1995. Although described here in terms of external consortium benchmarking, the process is generic and equally applicable in principle to all types of benchmarking.

The Juran 7-Step Benchmarking Process depicted in Figure 9.2 is divided into two phases. Phase 1 is a positioning analysis providing the benchmarker with a comprehensive study of the relative performance of all the benchmarking participants and a thorough consideration of the performance gaps to the top-performing superior or best-in-class organizations. The focus of phase 2 is upon learning from the phase 1 findings, adopting and adapting best practices, and developing improvement programs to implement the changes required. Each step in the process is described next.

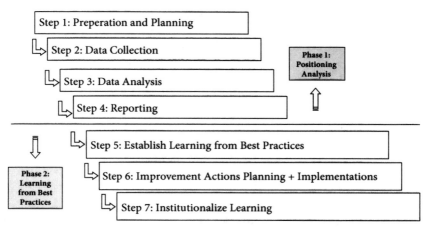

Figure 9.2 The Juran 7-Step Benchmarking Process.
(From Juran Institute, Inc., 2013)

Planning and Project Setup

Step 1 is to recognize the need for benchmarking, to clearly understand what is to be benchmarked and why, to determine the benchmarking methodology that is going to deliver the analysis required, and to identify who is to be benchmarked. A benchmarking project is no different from any other project. To succeed, thorough preparation and planning at the outset is essential. Often a business case will need to be made to justify the need for the benchmarking project.

Critical at this stage of designing the benchmarking is to clearly define the scope of the benchmarking envelope, what is to be benchmarked and what is to be excluded. The metrics to be used can then be agreed upon, and these too must be clearly and unambiguously defined to ensure comparability of data collected. Finally, the most appropriate vehicle for data collection must be determined.

Once the benchmarking topic has been well defined, the participants with whom the benchmarking will be conducted must be determined. As mentioned earlier, ideally those organizations that are known to be superior performers will be identified as participants in the benchmarking. However, the participants will be dependent upon the type of benchmarking being conducted as well as the way in which the participants are selected; but of course the ultimate aim is to benchmark with the recognized performance leaders.

During this initial planning step participants will also aim to do as follows:

- Identify and agree on the key performance indicators (KPIs) to be used to assess performance.
- Create a metrics model that clearly demonstrates the interrelationships among the metrics in use.
- Develop clear and unambiguous definitions for all the metrics used.
- Produce a data collection document as a vehicle for participants to collect and submit their data and conduct some initial validation of the data prior to submission.
- Agree on the project time schedule, milestones, and deadlines.

Data Collection and Normalization

Once the precise KPIs and associated definitions are identified, a method for collecting data from each participant must be developed. Commonly a data collection document is produced and issued to all participants to enable them to collect and submit their data. Data submissions are increasingly conducted online via secure Web portals. Proprietary spreadsheets are also frequently used because they are widely available (all participants are likely to have access to them); they are easy to use, have very powerful calculating capability, and can be tailored to provide automated functionality for validation and calculation. The data collection document must be designed to be easy and quick to populate by the user, to provide a suite of validation checks to maximize data quality and minimize errors.

Participant Support During the Benchmarking Process

To facilitate the data submission and validation process, it is a good idea to operate a help desk that is available during the entire project duration. This is often provided when third-party consultants are facilitating the benchmarking. The desk can provide professional advice on how to fill the data collection document and answer questions related to specific program-related matters (e.g., interpretation of definitions used). The objective is to provide a swift response to participants so as not to delay them in the data collection process. Of course, by providing clear and thorough guidance notes and a well-structured data collection document, the need for participants to seek help will be minimized. Nonetheless, a help desk can be an extremely valuable and essential source of benchmarking support, especially for newly established benchmarking programs. When required, a list of frequently asked questions (FAQs) can also be developed and forwarded to all participants.

Data Validation

Use of valid data is key to the success of any benchmarking program, where the adage "Garbage in, garbage out" has never been more appropriate. Incorrect or inaccurate data can easily result in misguided conclusions and inappropriate actions and can lead to the failure of any improvement program. Furthermore, endless rounds of clarification will lead to frustration by the participants and can delay the benchmarking process. Thus a high degree of emphasis should be placed on data validation. In Juran's benchmarking programs they adopt a two-phase approach, with the initial application of a suite of automated checks followed by a number of manual checks.

Automated checks are an integral part of the data collection document, which is designed in such a way that it is easy to populate and has a number of built-in validation checks, thereby maximizing data quality and minimizing errors. The built-in automated error checks aim to prevent the input of spurious data and enable users to conduct their own first-pass manual check of the data prior to submission. Thorough initial checking by the users themselves significantly reduces the time and effort required for subsequent validation.

Once the data are submitted, the facilitator should conduct a number of manual checks according to a rigorous data validation process. Juran typically employs a three-step process:

1. Data completeness
2. Data integrity
3. Data consistency

These checks should be carried out by an experienced individual who understands not only the benchmarking process but also the nature of the data being submitted and the interrelationship between different data points. All data are first checked to ensure that they are complete. A check of the integrity of the data should then be carried out by comparing different interrelated data to ensure that the expected relationship between these data is observed. Finally a range of intelligent triangulation checks can be conducted to further ensure consistency between data provided and any available historical data sets. Any anomalies should be raised one on one with the relevant participant to ensure that corrected data are provided. Where either there are a large number of apparent errors or participants are experiencing difficulty in obtaining the data required, a data clinic may be held, attended by participants and designed to clarify any confusion relating to the data required.

Data Normalization

The single biggest problem in any benchmarking exercise lies in how to compare benchmarked subjects on a like-for-like basis (i.e., how to compare apples with pears). In some circumstances the benchmarkers will be similar enough to enable direct comparisons of performance between them. However, more typically the subjects being benchmarked will all be different from one another, be they organizations as a whole, business units, different sites, different functional groups, business processes, or products. No two subjects will be identical, although the extent of difference between them will vary considerably depending upon what and who are being benchmarked. Thus to be able to compare differences in performance levels requires some intervention. Some form of data normalization is usually required to enable like comparisons to be made between what may be very different subjects. Without it, direct comparisons of performance are normally impossible and may lead to misinformed conclusions. Normalization can be made on the basis of a wide range of factors including scope, scale, contractual arrangements, regulatory requirements, and geographical and political differences.

Normalization is essentially the process of converting metrics into a form that enables their comparison on a like-for-like basis, accounting for all (or as much as possible) the variation between the benchmarking subjects. It is that the normalizing factor used be truly a driver for the performance being benchmarked. For example, in benchmarking the operating costs of the invoicing function of an organization, perhaps a suitable normalizing factor is the number of invoices raised. For example, the costs could be compared on a per invoice raised basis. However, perhaps some invoices are more complicated to produce than others (e.g., they may contain more line items or be for a higher total value that requires more checks before the invoice is raised), so this way of normalizing may not be appropriate after all.

The most common way of doing this is by looking at performance per unit or per hour. For example, if we are measuring the cost of manufacturing a motorcar, we might compare the cost per vehicle produced; or if we are looking at the time taken to treat a hospital patient with a given ailment, we might consider the number of patients examined per hour.

In some cases a simple measurement per unit is not sufficient to accommodate the variation observed between benchmarking subjects, and a more sophisticated approach has to be developed. In such cases the use of weighting factors that represent the variation of the different benchmarking subjects is often a very effective means of normalization. Weighting factors may be developed in relation to costs, time, and efficacy. An example of a highly effective weighting factor is the Juran Complexity Factor (JCF). The JCF was developed to enable like-for-like comparisons to be made between oil and gas production facilities of very different size and design. The normalizing factor takes into consideration the equipment present in the facility and the time it takes to operate and maintain this equipment under normal conditions. The JCF is then used to normalize all cost performance between facilities in the benchmarking. This enables organizations to directly benchmark their facilities with those of other organizations even though they may be very different in design and size.

Analysis and Identification of Best Practices

The aim of the analysis is to determine the findings from the data collected in the benchmarking in conjunction, where appropriate, with other pertinent data and information from a number of different sources including the public domain, the participants themselves, and any previous editions of the benchmarking study. The level of analysis will be dependent upon the scope and objectives agreed upon at the commencement of the benchmarking.

It is essential that the analysis be impartial and totally objective. It must also be aligned to the benchmarking objectives. And to be of value, it must indicate the benchmarker's strengths and weaknesses; determine, and where possible quantify, the gaps to the best performers; and identify as far as possible the reasons for these gaps. It is important that the metrics be considered collectively and not in isolation as the results from one metric may help to explain those of another. The strategies and working practices of each of the participants should also be explored and used to determine how they may influence performance.

The performance data and any normalization data streams are analyzed to compare participant performance and determine performance gaps. It is also important to consider the level of statistical testing of the data to ensure that comparisons being made are statistically significant and the conclusions drawn thereafter are valid.

Reasons for apparent differences in performance should be considered during the analyses. With multinational or global benchmarking studies, it is important to consider the impact that may be attributed to differences in geographical location. For example, when one is analyzing costs, it is clear that cost levels (e.g., salaries) in the West cannot be easily compared to those in the East, Russia, Africa, or Latin America. In addition, fluctuations in exchange rates between currencies can have a dramatic effect. Likewise different tax regimes, regulatory requirements, political policies, and cultural differences can all significantly influence performance.

Report Development

Once the analysis is complete, it must be reported to the benchmarking participants. The content of the report and the medium used for reporting were agreed to at the outset of the benchmarking exercise and in part are determined by the type of benchmarking being undertaken.

Unfortunately, many benchmarking exercises will stop at this point. But to maximize the value gained from benchmarking, organizations must go further to try to understand the practices that enable the leaders to attain their superior performance levels. This is the purpose of phase 2 of the Juran 7-Step Benchmarking Process.

Learning from Best Practices

A benchmarking program must go well beyond the comparison of performance data. The transfer of knowledge from the best practitioners to the other benchmarkers is critical. This is essential to maximize the effectiveness of knowledge transfer, which leads to highly successful change/process improvement programs. This can be achieved in a number of ways:

- Internal forums
- One-to-one benchmarking
- Best practice forums

Internal Forums

Organizations participating in the benchmarking should fully review the benchmarking report and consider the findings in detail. Thereafter many organizations find it beneficial to organize an internal forum attended by all parties within the organization affected by the benchmarking, to discuss the findings openly and determine first actions required to begin the process of gap closure and performance improvement. In cases where an organization may have numerous participants in the benchmarking (e.g., an organization may benchmark several different business units simultaneously), these internal forums can be an excellent platform for sharing knowledge among those business units. Juran's oil and gas experts have had the opportunity to attend and facilitate a series of internal knowledge exchange forums in the oil and gas industry (e.g., Pemex Mexico in 2004 and Qatar Petroleum and Saudi Aramco in 2008), where up to 500 members of staff from different departments within one organization gathered with the sole aim of sharing knowledge and best practices related to vital topics and processes.

One-to-One Benchmarking

It is common for organizations to benchmark one to one following participation in a group benchmarking study. Having identified superior performers in various areas of the benchmarking, organizations collaborate on a one-to-one basis to explore specific issues in greater detail, perhaps by on-site visits or further data exchange and analyses, to maximize the learning outcomes.

Best Practice Forums

This involves the sharing of best practices between top-performing organizations to the mutual benefit of all benchmarkers. Of course, when one is benchmarking with true competitors, the options for this may be limited and alternative approaches may be required to establish learning.

Once findings have been reported to all participating organizations, a best practice forum can be organized, attended by all participating benchmarking organizations. The best practitioners in each of the elements of the benchmarking model are asked to make presentations to the closed forum. Prior to the forum, all participants would be invited to submit any questions they might have for the best practitioners, which they would like to be addressed by the best practitioners in their presentations.

The objective of this forum is the identification of master class opportunities and the transfer of knowledge from the best practitioners to the other benchmarking partners. Best practitioners will present to their peers the "whys and the hows" of their best practices. The intention is that audience members will learn from these presentations, which will help them subsequently to formulate their own improvement programs. Participants should leave the best practice forums in a position to develop clear action plans for the implementation of improvement programs.

Improvement Action Planning and Implementation

Once the learning points have been ascertained, each organization needs to develop and communicate an action plan for changes required to realize improvements. Here the learning from the benchmarking will feed into the organization's strategic plan and be implemented using its performance improvement processes. But how can organizations translate benchmarking findings into action plans that will lead to performance improvement?

The output of a benchmarking exercise should become input for action planning. A typical output from a benchmarking exercise will include a series of performance gaps between a participant and the best practitioners in key business processes. Often organizations react with disbelief and denial when they are confronted with the performance gaps translated into monetary terms: Comments such as "It is impossible to save this much, these numbers cannot be right!" are commonplace. It is of paramount importance, for both the credibility of the benchmarking findings and the subsequent level of managerial buy-in, that one additional internal journey be embarked upon prior to moving to action planning. The organization needs to truly understand the performance gaps identified. Therefore it must eliminate any distorting elements from the gaps presented. To render the performance gaps actionable, the organization needs to break them down into controllable and noncontrollable gaps. Noncontrollable gaps are those relating to aspects of an organization's activities that are not under the direct control of that organization at the time. For instance, these could include start-up costs, one-off expenditures, extraordinary incidentals, regulations that one has to comply with, site-specific operational issues (e.g., climate, geography, topography), and geopolitical and safety-related costs.

An actionable performance gap should be free from noncontrollable elements (see Figure 9.3). This will allow management to do the following:

1. Assess a performance gap that they can relate to and therefore buy into.
2. Prioritize improvement areas and distinguish the vital few versus the useful many opportunities for improvement (Juran and Godfrey, 1995).
3. Allocate resources to fix the problems and bridge the gaps including an accountable project manager, budget, time frame, and targets. Managers and employees will then be empowered to get things done.
4. Put controls in place by embedding the requirement to action the improvements into managerial and individual employee target setting, compensation schemes, and business planning. This will enhance the chances of a successful implementation.

Thus, benchmarking findings will have been embedded into performance improvement action plans and integrated into routine business cycles, helping to ensure that resources are focused, individuals have bought into the process, and goals are achievable.

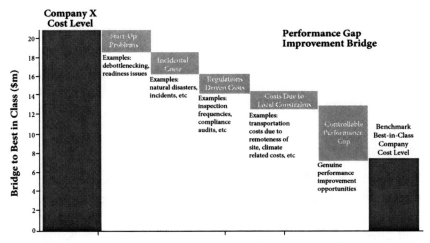

Figure 9.3 Performance bridge analysis from "gross to net performance gap." (From Juran Institute, Inc., 2013)

Institutionalizing Learning

Finally, the learning gained and the improvements to performance realized must be fully embedded to ensure all gains are rolled out throughout the organization and are sustained over time. Benchmarking may take place at the corporate, operational, or functional level within an organization, and it is important that each of these levels be linked via a cascading series of goals, interlinked to ensure systematic progress toward attaining the vision.

As improvement opportunities arise, they should be embedded into and replicated throughout participant organizations. Juran supports this step of the process by providing ongoing 24/7 support in the form of a members-only secure website accessible only by benchmarking participants. This provides a platform for information sharing and knowledge management well beyond the scope of a normal benchmarking program. It provides the opportunity for peers to learn from one another on a real-time basis "as and when required." In our increasingly rapidly changing business environment, it can be of great use to have direct access to the collective knowledge of specialists facing similar challenges.

Legal and Ethical Aspects of Benchmarking

The legality of benchmarking is governed by competition (antitrust) law and intellectual property law, and all benchmarkers must be aware of the legal and ethical implications of their benchmarking activity. While the ethos of benchmarking is the sharing of knowledge and information to the mutual benefit of all participants, organizations must not lose sight of the potential value of their corporate knowledge and therefore the necessity to adequately control its use.

The Benchmarking Code of Conduct

The Benchmarking Code of Conduct was first developed by the International Benchmarking Clearinghouse, a service of the American Productivity and Quality Center (APQC) in 1992 (see www.apqc.org). In 1996 a European version of the code was developed (see www.efqm.org), based upon the American version, to comply with European competition law. Neither document is legally binding, but they do lay down the principles for ethical and legal benchmarking. The main principles address legality, confidentiality, and information exchange, and all benchmarking programs should ensure that participants comply with these.

Confidentiality

Essential in all benchmarking studies is the requirement for some degree of confidentiality. The strictness of the level of confidentiality will be dependent upon the sensitivity of the subjects being benchmarked, the requirement to comply with competition law, and the degree of willingness by the participants to share data and information openly. It is clear that great care must be exercised when benchmarking prices. In many cases costs are considered an indicator of prices, and therefore strict confidentiality is normally also expected when comparing costs.

The degree of confidentiality exercised in benchmarking studies can vary enormously. At one end of the scale participants are totally unaware of whom they are benchmarking with, as the identities of the other participants are withheld and only the third-party facilitator is aware of who each participant is. Unfortunately a major drawback of such a strict level of confidentiality is that the learning potential is greatly reduced. If participants do not know the identity of the better performers in a benchmarking process, how can they possibly learn from the findings? The whole object of the benchmarking is lost, and the study becomes nothing more than a league table.

Therefore a more pragmatic approach is preferred whereby sensitive data (e.g., costs) can be anonymized whereas other less sensitive data can be shared more openly. And with the skillful support of a third-party facilitator the participants can still maximize the learning potential from the study.

Irrespective of the level of confidentiality and anonymity decided upon, it is essential that all parties in a benchmarking study, including the facilitator (be they a consultant or a participating organization), sign a confidentiality agreement. This agreement will be legally binding and will spell out how the data, information, and findings of the study will be shared, used, and disseminated by all parties.

Managing for Effective Benchmarking

For any benchmarking initiative to succeed, it must be managed effectively.

While senior managers are unlikely to be involved directly in conducting the benchmarking, they play a key role in ensuring it is executed successfully. Key roles of senior management are to do the following:

- Set benchmarking goals.
- Integrate benchmarking into the organization's strategic plan.
- Act as a role model.
- Establish the environment for change.
- Create the infrastructure for benchmarking.
- Monitor progress.

References

Anand, G., and R. Kodali. (2008). "Benchmarking the Benchmarking Models." *Benchmarking: An International Journal*, vol. 15, no. 3, pp. 257–291.

Godfrey, R. E. Hoogstoel, and E. G. Schilling (eds.), *Juran's Quality Handbook*, 5th ed. McGraw-Hill, New York.

Juran, J. M., and A. B. Godfrey. (1995). *Managerial Breakthrough*. Barnes & Noble, New York.

Kearns, D. T., and D. A. Nadler. (1993). *Prophets in the Dark—How Xerox Reinvented Itself and Beat Back the Japanese*. Harper Business, New York.

Wood, B. (2009). "7 Steps to Better Benchmarking." *Business Performance Management*, Penton Media Inc., October.

INDEX

accountability, BPM organizations, 39
action planning, benchmarking, 262–263
activities
   BPM process analysis, 225–226
   in process development, 122
actual use, 92, 107
adaptability
   business process, 212, 226
   in Juran Transformation Model, 29
   process measurement, 223
adaptability, breakthroughs in
   adaptive cycle and its prerequisites, 50–52
   data from external environment, 52–56
   data from internal environment, 52
   overview of, 49–50
analysis, BPM process, 225–226
anonymity, of benchmarking studies, 265
assessment, BPM process, 234
audits
   business, 84–87
   quality control system, 209
automated checks, benchmarking, 257

balanced business scorecards, 69, 85, 215
Baldrige Index, 60
benchmarking
   analysis and best practices, 259–260
   case study, 267–269
   classification based on data/information sources, 247–253
   classification based on subject matter, 242–247
   classifications of, overview, 241–242
   data collection/normalization, 256–259
   goal statement for, 98
   improvement action planning/implementation, 262–263
   institutionalizing learning, 264
   learning from best practices, 260–262
   legal/ethical aspects, 264–265
   managing, 266
   objectives of, 240
   planning and project setup, 255
   process of, 253–254
   as quality initiative strategy, 12
   reasons for, 241
   redesigning process with, 227
   report development, 260
   understanding, 237–240
Benchmarking Code of Conduct, 264–265
best practices, benchmarking, 12, 52, 239, 259–262
bite-sized breakthrough projects, 163
BPM. *See* business process management (BPM)
breakthroughs
   in adaptability, 48–56
   in culture, 42–48
   in current performance, 41–42
   defined, 4
   in management, 35–37
   in organizational structure, 37–40

breakthroughs (*cont.*)
   strategic planning and, 59, 62
   transformational change and, 30–35
   transforming culture for, 28–30
breakthroughs in current performance
   business plan goals, 156–158
   COPQ vs. cost reduction, 144–146
   driving bottom-line performance, 146–148
   executive "quality council," 152–155
   mobilizing for, 151–152
   models/methods, 132–142
   nomination/selection process, 158–165
   overview of, 129
   results, 148–151
   review progress, 172–173
   team organization, 165–172
   universal sequence for, 129–132
   upper management approval/participation for, 142–144
breakthroughs in current performance, models/methods
   disillusioned by failures, 141
   employee apprehensions, 142
   expenditure of effort required, 137–138
   illusion of delegation, 141
   inhibitors, 141
   lessons learned, 133–134
   major gains from few vital projects, 140–141
   project backlog, 137
   project by project, 136–137
   rate of breakthrough, 134–136
   ROI, 138–140
business excellence models, benchmarking, 244
business plans
   annual goals, 70–72
   breakthrough goals, 156–158
   strategic planning of, 57–58

business process improvement, BPM, 234
business-process managed organizations, 39
business process management (BPM)
   combining with technology, 234–235
   creating readiness for change, 231–234
   deploying, 215–216
   life cycle, 212
   methodology, 214–217
   operational management phase, 214
   organizing for, 216–217
   reasons for, 211–214
   simplifying macro processes with, 211
   transfer phase, 214, 229–231
business process management (BPM), planning phase
   analyzing process, 225–226
   control points, 224–225
   creating new process plan, 229
   customer needs/mapping current state, 219–221
   defined, 214
   defining current process, 218–220
   overview of, 217–218
   process measurements, 221–224
   redesigning process, 226–229
business unit or site location benchmarking, 243

Camp's 10-step benchmarking process, 253–254
case study, benchmarking, 267–269
certification, product conformance, 201
change
   belief in continuous adaptive, 48
   BPM and readiness for, 231–234
   breakthrough and transformational, 30–35
   breakthrough apprehensions of employees, 154–155

breakthroughs in adaptability,
    49–56
  as corrective action, 203–205
  goal statement for, 99
  of norms and behavior, 43–45
  resistance to, 45–48
chargebacks, sparing use of, 203
classifications, benchmarking
  by data and information sources,
    247–253
  by subject matter, 241–247
cloning breakthrough projects,
    163–164
common causes of variation, 192,
    194–195
competition, in twentieth century,
    10–11
competitive benchmarking, 245–246
competitive performance goals, 70, 72
competitive quality evaluation
  business audits, 85–87
  costs of poor quality, 83
  overview of, 81
  performance of business processes,
    83–84
  performance on improvement,
    81–83
  product and process failure, 83
  scorecard, 84–85
complaint handling, 108–109
compliance, 175–177
compliant process. See control process
conduct
  Benchmarking Code of Conduct,
    264–265
  code of, 47
  disposing of unfit products,
    202–203
confidentiality, benchmarking, 265
consistency, of benchmarking data, 257
consortium benchmarking, 250–251
constraints, goal statement for, 98
continuous improvement
  benchmarking for, 239
  continual breakthrough for, 31–32

norms for cultural transformation,
    47
  as quality initiative strategy, 12, 19
continuous perpetual cycle, 50–51
control points
  BPM process, 221–222, 224–225,
    234
  corrective action for out-of control,
    194–195
control process
  audits, 209
  definition of, 175–177
  developing, 126–127
  feedback loop, 178–179
  feedback loop elements, 180–187
  planning for, 190–192
  policy manual and, 208
  process conformance, 192–199
  product conformance, 199–205
  pyramid of control, 187–190
  relation to quality assurance,
    177–178
  statistical methods in, 205–207
  tasks for leaders, 209–210
controllable gaps in performance,
    benchmarking, 262–263
controls
  creating quality, 19, 23–24
  financial, 23
  Juran Trilogy diagram, 24–25
  as universal, 16
corrective action, 203–205, 209
cost
  BPM process measurement, 224
  breakthroughs in current
    performance, 41–42
  COPQ vs. reduction of, 144–146
  future of BPM combined with
    technology, 235
  of high quality, 2–4
  lowering by strategic planning, 61
  percentage of deficiencies vs.,
    162–163
  of producing goods and services,
    19–20

cost (cont.)
  quality and higher market share, 6–8
cost of poor quality (COPQ)
  annual goals for reducing, 70–72
  calculating, 22
  cost reduction vs., 144–146
  driving bottom-line performance, 146–148
  Juran Trilogy diagram and, 24–25
  measuring competitive quality, 83
  savings from reducing, 31
  securing need for breakthrough from upper management, 143–144
  tools and techniques for studying, 53
critical success factor, deploying BPM, 215
cross-functional business processes
  BPM, 213, 215–216
  BPM redesign, 228, 231
  breakthroughs in organizational structure, 37–38
  deployment process, 76
  Quality by Design project, 101
cross-functional teams, 38–39, 101
cultural patterns, 28–30, 45–48
culture
  breakthroughs in, 42–48
  creating sustainable. *See* leadership
  customer need for, 106
  Juran Transformation Model, 29–30
current performance. *See also* breakthroughs in current performance, 29–30, 52
current state of organization
  creating readiness for change, 231
  mapping in BPM, 219–221
  in review process, 80, 253
  in vision statement, 67
customer loyalty and retention
  annual goals for, 60–61, 71
  studies gauging, 54–55

customer needs
  analyzing and prioritizing, 110
  to be kept informed, 109
  in BPM planning phase, 219
  collecting list of in their language, 109
  cultural needs, 106
  human safety needs, 107
  overview of, 103–105
  perceived needs, 106
  product/service features and goals for, 114
  promptness of service needs, 108
  Quality by Design spreadsheets, 110
  related to failures, 108–109
  stated vs. real needs, 105–106
  traceable to unintended use, 107
  translating and measuring, 111
  user-friendly needs, 107
customer satisfaction
  absence of failures and, 91
  annual goals for, 71
  goal statement for, 97
  presence of features creating, 90
  quality initiative strategies, 12
  in quality of goods/services, 18–20
  as strategic planning focus, 58–61
customers
  adaptive cycle prerequisites, 52
  changes in habits of, 8
  cultural transformation of, 47
  as focus, 40
  identifying, 101–104
  in twentieth century, 10
  vision statement for, 66–68

data
  classifying benchmarking by, 247–253
  collecting benchmarking, 256
  measuring quality of, 52
  normalization of benchmarking, 256, 258–259
  quality. *See* adaptability, breakthroughs in

validation of benchmarking, 256–257
database benchmarking, 248
decision making
　disposing of unfit products, 202–203
　on fitness for purpose, 201–203
　leadership avoiding quality control, 209
　process conformance, 196–198
　product conformance, 199–201
　statistical methods in control, 207
decomposition, BPM process analysis, 225
delegation, illusion of management by, 141
Deming Prize, 86
departments, horizontal flow through functional, 212–213
deployment plan. *See also* strategic planning (SP), and deployment
　BPM, 215–216
　defining, 65
　new process, 232
Design for Six Sigma (DFSS), 54
design gaps, 94
design reviews, 120–121
DFSS (Design for Six Sigma), 54
diagnosis, breakthrough, 171–172
dictatorship, leadership vs., 37
disasters, rapid adaptive action avoiding, 49
documentation, final process design, 121–122, 125–126

effectiveness, BPM, 212, 223, 225–226
efficiency, BPM, 212, 223, 225–226
EFQM (European Foundation for Quality Management), 59–60, 85, 244
elephant-sized breakthrough projects, 163
employees
　achieving high performance, 39–40
　active participation in change, 34

　breakthrough apprehensions of, 142, 154–155
　empowering for quality initiative, 12
　self-control and controllability of, 197–198
　vision statement and, 66–68
ethics
　benchmarking, 264–265
　code of, 47
　strategic planning, 65
European Foundation for Quality Management (EFQM), 59–60, 85, 244
executive council, 73–74
executive owner, BPM, 216–217
executive "quality council," breakthroughs, 152–155
external benchmarking, 232, 245–247
external environment, 52–56, 101–104

facilitators (Black Belts), breakthroughs, 166–170
failures
　breakthroughs in current performance, 41–42
　breakthroughs to reduce excess, 129–131
　as competitive quality metric, 83
　customer needs related to, 108–109
　disillusionment of breakthrough, 141
　effect on cost and income, 21–22
　history of feature, 7–9
　Juran Trilogy diagram and, 24–26
　managing for quality, 19
　quality gaps, 93–94
　removing for quality improvement, 90–94
　strategic deployment risks, 62–63
features
　customer need for user-friendly, 107
　detailing goals for, 119–120
　developing process for, 122–126

272 | Index

features (*cont.*)
    fitness for use as presence of, 91–92
    grouping related customer needs, 114
    high-level goals, 118–119
    history of failures, 7–9
    managing quality of, 19–20
    methods for identifying, 114–118
    optimizing goals, 120–121
    overview of, 112–114
    product conformance and, 199–200
    as purpose of Quality by Design, 90–91
    setting /publishing final design of, 121–122
feedback loop
    closing with corrective action, 203–204
    compliant process and, 178–179
    elements of, 180–187
    merits of self-inspection, 200
financial processes, 23
firefighting, 203–204
fitness for purpose. *See* fitness for use
fitness for use
    evolution to, 17–20
    managing for quality, 16–17
    as presence of features and absence of failures, 91–92
    product conformance and. *See* product conformance
    quality as, 3
flow diagram, 103–104
flowchart, BPM, 219–222, 225
forums, benchmarking, 261–262
function-based organizations, breakthroughs in, 38
functional benchmarking, 242–243
functional change, 35
functions, breakthroughs in, 32–35, 37–38

gaps in performance, benchmarking, 240, 262–263
generic benchmarking, 244

goals
    aligning with strategic plan. *See* strategic planning (SP), and deployment
    annual, 64–65, 70–73
    in BPM planning phase, 217
    for business plan breakthrough, 156–158
    long-term, 59
    for process features, 122–126
    for product/service features, 118–122
    for project/design, 95–100
goods. *See* products

handoffs, eliminating in BPM, 228
help desk, during benchmarking, 256
historical performance, goal statement for, 99
horizontal deployment, transfer phase in BPM, 232

implementation problems, BPM addressing, 229–230
improvement
    action planning for, 262–263
    chronic waste as opportunity for, 25
    financial, 23
    process, 12
    programs for breakthrough, 41–42
    quality, 24
independent checkers, 198–199
industrial tourism, 238
information
    adaptive cycle and, 50, 52–56
    evaluating for breakthroughs in adaptability, 49
Information Quality Council, 49
information technology (IT). *See* technology
infrastructure
    breakthrough, 164–165
    implementation problems in BPM, 231

inhibitors, breakthrough, 141
initiatives, strategic planning, 65
inspector, in process conformance decisions, 198–199
institutionalizing
 breakthrough, 172
 of learning in benchmarking, 264
integrity, benchmarking data, 257
intelligence findings. See adaptability, breakthroughs in
intended use
 defined, 91–92
 developing process features, 123
 needs traceable to, 107
 vs. unintended use, 100
interdepartmental cooperation, strategic planning, 62
internal benchmarking, 245–246
internal customers, 102–103, 106–107
internal forums, benchmarking, 260–261
IT (information technology). See technology

Japanese or Toyota quality, 17–18
Juran 7-Step Benchmarking Process, 254
Juran Complexity Factor (JCF), 259
Juran Quality by Design model
 customer needs, 103–112
 defined, 90
 examples of, 90
 identify customers, 101–103
 overview of, 90–93
 process controls and transfer to operations, 126–127
 process features, 122–126
 product design spreadsheet, 112–113
 product or service features, 112–122
 project and design goals, 95–100
 project team, 101
 steps, 95

Juran Transformation Model
 breakthroughs and transformational change, 30–35
 breakthroughs in adaptability, 48–56
 breakthroughs in culture, 42–48
 breakthroughs in current performance, 41–42
 breakthroughs in management, 35–37
 breakthroughs in organizational structure, 37–40
 culture defined, 27–30
 defined, 27
Juran Trilogy. See also product innovation, 16, 23–26

key objectives, strategic planning, 64
key performance indicators (KPIs)
 defining in strategic planning, 65–66
 measuring progress, 77–80
 planning benchmarking project, 255
 on scorecards, 84
key terms. See terminology

leadership. See also Juran Transformation Model; management
 adaptive cycle prerequisites, 51
 avoiding quality control decisions, 209
 of breakthrough team, 165–172
 breakthroughs in, 35–37
 challenges of implementing change, 35
 dictatorship vs., 37
 major elements of, 36
 quality initiative strategies, 12
 role in strategic planning, 73–81
Lean Six Sigma, 21, 41–42, 58
legality of benchmarking, 264–265
limit lines, Shewhart control chart, 193–197

list, of customer needs, 109
long-term goals, strategic planning, 64
long-term planning, benchmarking, 252–253
long time cycles, breakthroughs in current performance, 41–42

macro processes, simplifying. *See* business process management (BPM)
Malcolm Baldrige National Quality Award
   as benchmarking model, 244
   first winner of U.S., 18
   merits of, 5
   strategic planning integral to, 58–60
management. *See also* leadership
   approval for breakthrough from upper, 142–144
   breakthroughs in, 35–37
   for effective benchmarking, 266
   executive "quality council" responsibilities, 152–155
   mobilizing for breakthrough, 151–152
   for quality, 16–20
management-controllable products, 198
manual checks, benchmarking, 257
mapping current state, 219–221
market
   adaptive cycle prerequisites, 52
   building quality leadership, 5–6
   goal statement, 97–98
   quality/growth in share of, 6–8
   research, 55
measurement (metrics)
   BPM, 212
   BPM planning phase, 221–225
   competitive quality. *See* competitive quality evaluation
   of customer needs, 111
   goal statement for success, 99
   of internal affairs, 53

normalization of data by converting, 258–259
of performance with KPIs, 77–80
planning benchmarking project, 255
progress review, 80–82
members, breakthrough team, 166–170
methodology, BPM, 214–217
mission, organizational, 64–66, 68–69
mission statement, 217
mobilizing, for breakthrough, 151–152
monopolies, 72
motivation, for setting goals, 36–37
multifunctional teams
   assuring compliant processes, 191, 207
   in BPM, 217
   for design, 121
   planning breakthrough infrastructure, 164
   planning Quality by Design project, 103
   process and quality improvement with, 12
   vital few breakthrough projects by, 140–141, 162

nomination, breakthrough projects, 158–160
noncompetitive benchmarking, 245, 247
noncontrollable gaps in performance, benchmarking, 262–263
normalization, benchmarking data, 256, 258–260
norms, cultural
   achieving cultural transformation, 47–48
   acquiring in workplace, 43
   policies, 48
   reinforcing desired changes, 43–45
   resistance to changing, 45–48

one-to-one benchmarking, 250, 261
open systems, 34–35

Operational Excellence (OpEx), 59
operational management phase, BPM, 214, 233–234
operations gaps, 94
optimization of design, 120–121, 125
Organizational Effectiveness programs, 20–21
organizational structure
    adaptive cycle and, 51
    breakthroughs in, 37–40
    Juran Transformation Model, 29–30
    rapid adaptive action to avoid disasters, 49
organizations
    breakthroughs essential for vitality of, 30–32
    implementation problems in BPM, 231
    readiness for change, 232
owner, process
    accountability assigned to, 39
    BPM analysis, 226–227
    BPM methodology, 214
    BPM process measurements, 221, 223
    BPM qualifications for, 211
    BPM, readiness for change, 231–234
    BPM responsibilities, 216–219
    deploying goals, 75–76
    employees acting as, 39–40
    managing cross-functional processes, 37
    of multifunctional processes, 77
    quality council for, 153

Pareto principle, 157–158, 163
people
    as benchmarking model, 232
    implementation problems in BPM, 231
perceived needs, of customers, 106
performance
    achieving high, 39–40
    benchmarking determining superior, 240
    breakthroughs in current. *See* breakthroughs in current performance
    causes of problems, 52
    cultural patterns determining, 28
    goal statement for, 97
    goals for improvement of, 70
    measuring competitive quality of, 81–84
    measuring with KPIs, 77–80
    profitability of strategic planning, 60
    quality initiative strategy for, 13
    setting goals based on historical, 72–73
Performance Excellence, 21
performance excellence department, 155
planning
    BPM. *See* business process management (BPM), planning phase
    for control, 190–192
    financial, 23
    implementation action in BPM, 232
    Juran Trilogy diagram for, 24–25
    long- and short-term, for benchmarking, 252–253
    for quality, 16, 23–24
policies
    cultural norms for quality, 48
    defining in strategic planning, 65
    deployment of, 58
    goal statement for new products, 99–100
    implementation problems in BPM, 231
policy manual, 208, 210
predictable change, breakthroughs and, 30
prerequisites, adaptive cycle, 50–52
president's audit, 87
prioritizing customer needs, 110
priority deployment, transfer phase in BPM, 232

procedures, policy manual, 208
Process Analysis Summary Report, 226
process-based organizations, 39
process benchmarking, 243
process conformance
    causes of variation, 192
    effects on decision of, 198–199
    points outside of control limits, 194–195
    points within control limits, 193–194
    self-control and controllability, 197–198
    Shewhart control chart, 192–193
    statistical control limits and tolerances, 195–197
process control, BPM, 234
process gaps, 94
process map, deploying BPM, 216
process owner. *See* owner, process
process plan, creating, 229–230
processes. *See also* business process management (BPM)
    annual goals for performance of business, 71
    benchmarking, 253–254
    breakthroughs based on managed business, 39
    breakthroughs in cross-functional, 37–38
    capabilities of key repetitive, 52
    developing controls/transfer to operations, 126–127
    developing features, 122–126
    measuring competitive quality of business, 83–84
    performance of key repetitive, 52
    simplifying macro. *See* business process management (BPM)
product conformance
    corrective action, 203–205
    decision making, 199–200
    diagnosing sporadic change, 204–205
    disposition of unfit product, 202–203
    fitness for purpose decision, 201–202
    overview of, 199–200
    self-inspection, 200–201
product innovation
    designing innovative products, 89–90
    overview of, 89
    Quality by Design model for. *See* Juran Quality by Design model
    Quality by Design problem, 93–94
products
    benchmarking and designing new, 251
    managing for quality, 21–24
    meeting or exceeding customer requirements, 19–20
    policies for quality, 48
    quality performance goals for, 70, 72
    salability of. *See* competitive quality evaluation
project benchmarking, 243–244
projects
    backlog of breakthrough, 137
    defining in strategic planning, 65
    identification of, 95–96
    nomination/selection process for breakthroughs, 158–165
    planning/setting up benchmarking, 255
promotions, 44–45
promptness, customer service need, 108
publishing
    final process features and goals, 125–126
    final product features and goals, 121–122
pyramid of control, 187–190

quality
    changes in customer habits and, 8
    by design. *See* Juran Quality by Design model

establishing goals, 60–61
freedom from failures, 19
meanings of, 90
meeting customer needs, 19
norms achieving cultural
 transformation, 47
strategic planning for. *See* strategic
 planning (SP), and deployment
quality assurance, in compliant
 process, 177–178
quality directors, 155
quality, embracing
 building market quality leadership,
  5–6
 earnings and stock market, 5
 growth in market share, 6–8
 impact on revenue and costs, 4
 lessons learned, 11–13
 meanings of, 3
 sustainable business results of
  superior goods, 1–4
 in twentieth century, 8–11
 in twenty-first century, 11
quality gaps, 93–94
Quality Planning, 54

rate of breakthrough, 134–136
RCA (root-cause analysis), quality
 control and, 200
real needs of customers, 105–106
redesigning process, BPM, 226–229
regulations, implementation in BPM,
 231
repetitive-use quality control systems,
 208
replication of breakthrough projects,
 163–164
reports
 benchmarking, 260
 internal affairs, 53
 managers disputing reliability of,
  50
 requiring universal participation,
  61
 on scorecard, 84–85

research, ROI for breakthrough
 improvement, 138–140
resistance to change, 45–48
results, breakthrough improvement,
 148
return on Investment (ROI),
 breakthrough projects, 138–140,
  148–151, 162
revenue
 annual goals for increasing, 72
 BPM process measurement, 224
 effect of features on, 21
 impact of quality on, 4–8
 strategic planning increasing, 61
review
 breakthrough process, 172–173
 periodic process assessment in
  BPM, 234
 plan for progress, 80–82
rewards
 establishing annual goals, 61
 how norms are changed, 44–45
risks
 of statistical process control,
  206–207
 of strategic deployment, 62–63
role(s)
 of leadership in strategic planning,
  73–81
 in transformational change, 32–35
root-cause analysis (RCA), quality
 control and, 200

safety needs, of customers, 107
satisfaction. *See* customer satisfaction
SC (special-contract), process in
 BPM, 219–220, 222
science, trends affecting your
 organization, 55
scorecard
 balanced business, 69
 defining in strategic planning,
  65–66
 deploying BPM with balanced
  business, 215

scorecard (*cont.*)
    gathering information about internal affairs, 53
    measuring competitive quality, 84–85
selection process, breakthrough projects
    cost figures, 162
    cost vs. percentage of deficiencies, 162
    elephant-sized/bite-sized projects, 163
    model of infrastructure, 164–165
    nomination, 158–160
    overview of, 160–161
    replication and cloning, 163–164
    vital few and useful many, 161–162
self-assessment benchmarking, 249–250
self-control
    in process conformance, 197–198
    self-inspection requiring, 201
    self-inspection vs., 200
self-inspection, product conformance, 200–201
services
    customer need for prompt, 108
    developing features of. *See* features
    dimensions of quality for, 19
    managing for quality, 21–24
    meeting or exceeding customer requirements, 19–20
    in product innovation. *See* Juran Quality by Design model
Shewhart control chart, 192–193
short-term planning, benchmarking, 252–253
six C's, today's business environment, 211
Six Sigma
    breakthroughs in current performance problems, 41–42
    breakthroughs in current performance using, 133
    defined, 4

gathering information on internal affairs, 53
merits of, 5
replacing Total Quality Management, 20–21
ROI for breakthrough improvement, 139–140
strategic planning integral to, 58–59
universal of, 16
U.S. organizations using, 18
Six Sigma DMAIC, 133
socialization, in workplace, 43–45
society
    2008 economic crisis and, 55–56
    rewarding members for conformance, 28
    trends affecting your organization, 55
    vision statement benefits to, 66–68
    your organization as a, 27
SPC (statistical process control), 196, 205–207
special causes of variation, 192, 194–195
special-contract (SC), process in BPM, 219–220, 222
sporadic change, diagnosing, 204–205
spreadsheets
    benchmarking validation, 256
    product design, 112–113
    Quality by Design, 110–111
stakeholders, organizing for BPM, 216–217
standards, policy manual for quality control, 208
stated needs, of customers, 105–106
statistical control limits, 195–197
statistical limit lines, 192–193
statistical process control (SPC), 196, 205–207
strategic planning (SP), and deployment
    agreeing on mission, 68–69
    aligning quality goals with, 57
    benchmarking input for, 241

benefits of, 57–58
competitive quality. *See*
  competitive quality evaluation
defining, 58–60
deployment benefits, 61–62
deployment risks, 62–63
deployment to whom, 76–77
developing annual goals, 70–73
establishing vision, 66–68
launching, 63–66
leadership role in, 73–81
Malcolm Baldrige National Quality Award, 58
measuring performance with KPIs, 77–80
model for, 59
quality and customer loyalty goals, 60–61
reviewing progress, 80–81
subdividing and deploying goals, 74–76
tree diagram for, 76–77
subgoals, deployment, 74–77
subject matter, classifying benchmarking, 242–247
subprocesses, BPM process analysis, 225
subsystems, in transformational change, 32–35
survey benchmarking, 249
surveys, 53–54
sustainability. *See* adaptability, breakthroughs in; leadership
systems thinking, transformational change and, 32–35

tasks
  BPM process analysis, 225–226
  control process leadership, 209–210
  process development, 122
teams
  cross-functional, 38–39, 101
  mobilizing for breakthrough, 152
  multifunctional. *See*
    multifunctional teams

organizing breakthrough, 165–172
organizing for BPM, 216–217
technology
  BPM critical to, 212, 234–235
  consumer movement in twentieth century, 10
  explosive growth/threats of, 9
  goal statement for, 98
  implementation problems in BPM, 231
  trends affecting your organization, 55
terminology
  annual goals, 61
  benchmarking, 239
  breakthrough, 129, 137, 171
  special and common causes of variation, 192
  strategic deployment, 64–66
  translating customer needs into "our" language, 111–112
threats. *See* adaptability, breakthroughs in
time, Shewhart control chart, 193
tolerances, statistical control limits and, 195–197
TQM (Total Quality Management), 20–21, 58
training
  annual goals for, 61
  breakthrough, 138
  as breakthrough facilitator, 169–170
  as quality initiative strategy, 12
  workforce for product conformance decisions, 201
transfer phase
  BPM, 214, 229–231
  product plan to operation, 126–127
transformational change
  continual breakthroughs for, 30–32
  norms in achieving, 47–48
  systems thinking and, 32–35
translation, of customer needs into our language, 111
tree diagram, 76–77

trial deployment, transfer phase in BPM, 232
The Triple Role, 33
TRIZ, 54
troubleshooting, in product conformance, 203–204
twentieth century, changes in, 8–11
twenty-first century, quality in, 11
two-tier ownership, BPM, 216

understanding gaps, 94
unfit products, disposal of, 202–203
unintended use
  customer needs traceable to, 107
  intended vs., 100
universal quality management
  concept of universals, 15–16
  effect of failure on cost, 22
  effect of failure on revenue, 21
  effect of features on revenue, 21
  financial analogy of, 22–24
  Juran Trilogy diagram, 24–26
  managing for quality, 16–20
  Organizational Effectiveness programs, 20–21
  overview of, 15
universal sequence, breakthroughs, 129–132
universals, 15–16
use
  fitness for. See fitness for use
  intended. See intended use
  unintended, 100, 107
user-friendly features, 107

validation
  of benchmarking data, 256–257
  transfer of product plan to operation, 127
value statement, 48
values
  in strategic planning, 65
  workplace shared, 27–28

vanguard companies, 2
variation, process conformance, 192, 194–195
vertical deployment, BPM transfer phase, 232
vision, organizational
  benefits of, 57
  creating readiness for change, 231
  definitions of, 64
  deploying, 64–66
  establishing, 22, 66–68
  mission statement vs., 68–69
  pitfalls, 68
voice of customer, 61

warranties, 108
waste
  annual goals for reducing, 70–71
  breakthroughs reducing chronic. See breakthroughs in current performance
  carryover of failure-prone features, 7–8
  discovering, removing and preventing causes, 31
  ROI in reduction of chronic, 138–140
weighting factors, normalization, 259
workflow problems, in BPM, 231
workforce
  breakthroughs in, 36–37
  process conformance decisions, 198–199
  product conformance decisions, 200–201
workplace, 27–28, 43
world-class companies, 2

Xerox Corp., 237–238

CPSIA information can be obtained
at www.ICGtesting.com
Printed in the USA
LVOW10*0247041017
551100LV00007B/60/P